MEN AND MOMENTS
IN THE HISTORY
OF SCIENCE

MEN AND MOMENTS
IN THE HISTORY
OF SCIENCE

Edited by Herbert M. Evans

GREENWOOD PRESS, PUBLISHERS
NEW YORK

PUBLISHER'S NOTE

The essays in this book are presented as part of the celebration of the twenty-fifth anniversary of the History of Science Dinner Club, which was founded in 1932 on the Berkeley campus of the University of California. The impact of this club on its field of study can readily be estimated from the quality of the present essays.

The lively activity of the club reflects the creative efforts of its founder, Dr. Herbert M. Evans, who assembled these papers and saw them through the press. To him the club and the contributors to this volume owe a debt of gratitude.

CONTENTS

Contents

MEN AND MOMENTS
IN THE HISTORY
OF SCIENCE

ONTOGENETIC AND OTHER DEVELOPMENTAL PARALLELS TO THE HISTORY OF SCIENCE

BY EGON BRUNSWIK

It has been argued that in any natural development higher order products must pass through a necessary sequence of more primitive, preliminary stages. * The famed biogenetic law, conceiving of ontogenesis as a recapitulation of phylogenesis, is but one of a variety of possible applications of this general idea. Among other things, it has suffered from the danger of overspecification in that it became difficult to decide which particulars of phylogenesis were essential and which were incidental and thus dispensable. This also holds for the first major attempt to apply the biogenetic law to psychology, undertaken by the American psychologist G. Stanley Hall in his book on adolescence in 1904.[1] Hall assumed that children pass through the stages of hunting, building, and so forth, as they mature, in alleged recapitulation of the history of mankind. This theory, based on faulty anthropology and stated in overly concrete categories without sufficient evidence, has not stood the test of time.

A more promising way to establish development parallel-

---

*This is a somewhat revised version of a paper read before the History of Science Dinner Club at Berkeley in 1939. While the essentials of the original presentation were left intact, it has been brought up to date in certain respects, primarily so far as the writer's own views relating to the subject are concerned.

isms in psychology is by concentrating on certain more broadly conceived, fundamental modes of approach or organization,
principles so basic and abstract as to be unaffected by situa-
tionally determined details of the adjustmental process. Aside
from some psychologically noncommittal early formulations of
Herbert Spencer and of Wundt centering about "homogeneity-
heterogeneity" or "differentiation, " most of the attempts along
this line came predominantly from the European continent within the last thirty-odd years. Some of these, such as notably
that of Heinz Werner, [2] deal mainly with ontogeny, that is,
child development, in relation to phylogeny, the latter in the
form of comparative material from both *primitive cultures* and
animal experimentation. From the standpoint of our topic,
the history of science is a part of phylogeny, representing as
it does a particularly lofty aspect of the general cultural development. The idea that children's theories may more or less
spontaneously move along paths previously traveled by the
great creators of scientific ideas has been championed especially by Piaget[3] and his collaborators at Geneva. We shall
refer to some examples of his work below.

Perhaps even more daring than the search for extraneous
counterparts of ontogeny is the execution of the idea that the
individual being contains a variety of developmental stages
within himself. In modern psychology this *intraindividual genetic* approach has taken three major forms.

The first is incorporated in the conception of the adjustmental makeup of organisms as a *simultaneous hierarchy* of functions operating at different developmental levels. In the present context it may suffice to point to various efforts, in essence dating back to Helmholtz, of describing "perception" as
a primitive mode of "thinking, " more specifically of inductive
reasoning. Some relation to the notions of the romantic natural philosophers of the early nineteenth century, notably
Oken, concerning the simultaneous hierarchy of organs of different developmental status within the body of higher animals
may here be discerned. [4] In application to ontogeny it has been
pointed out that the thinking of children is in many ways per-
ceptionlike; as regards our present topic, we shall bring up
cases in which early scientific theories show structural resemblance to certain specific patterns found in perception.

The second form of intraorganismic developmental comparison is given by the fact that the developmental level at which an organism operates seems to some extent to be determined by the nature of the demands of the stimulus situation; thus we have what may be called a task-induced hierarchy. For example, a simpler task may be handled by a more advanced function, while a more difficult or complex one of the same general kind may involve recourse, or "regression, " to a lower developmental level. Among others, Werner[5] has advanced certain ideas along this line.

A third application of the comparative approach within the same individual is given by attempts to subsume certain relatively rapid process chains under the genetic point of view so that they appear as a *telescoping of the developmental process*. There are two subvarieties of this third case.

a. One concerns *practice-induced changes*. These are changes occurring in the course of what may broadly be labeled practice or familiarization, in contradistinction to learning proper. A practiced function or person is in certain ways comparable to a function or person at a higher developmental level. One of the most notable attempts to establish exercise of a mental function as an effective substitute for general level of development, the latter involving both the natural level of intelligence and impairments through brain damage, has recently come from the American psychologist, Harry Harlow.[6] The examples presented in this paper will be from other contexts, however.

b. The second subvariety of intraindividually telescoped development concerns what Sander[7] and others in the Leipzig school of developmental psychology have called "actualgenesis, " as most drastically exemplified by the almost instantaneous growth of perceptual actuality as we open our eyes to a new scene. Sander brings forth evidence for his belief that every single cognitive act runs the gamut of an intrinsic genetic hierarchy, so that actualgenesis becomes a third partner to ontogenesis and phylogenesis within an expanded biogenetic scheme.

Let us turn to some concrete illustrations. Some of the examples cited will have ramifications with respect to more than one of the genetic parallelisms listed above.

Developmental Aspects of Criticism and the Constructive
Interpretation of Error

We shall begin with experiments on the development of criticism in children undertaken as a University of Vienna doctoral
dissertation by Anna B. Brind some twenty-five years ago,
under the direction of this writer. [8] There are several reasons
for choosing this study as our first example. From the beginning its plan was geared to certain problems in the philosophy of science, specifically the ways in which new propositions
are rejected or accepted. It has been abundantly observed by
students or contemplative onlookers of the history of science
that novelty tends at first to be met by stunned and rather in-
discriminate rejection, followed by more thoughtful and specific criticism--a pattern that makes the history of science the
drama it is. The very generality and atmospheric character
of the attitudes involved make it possible to concentrate on
formal modes of development rather than on its content, as we
said at the beginning that we should. A further advantage of
this study is a design which permits us to bring into the picture both stimulus-induced developmental levels and intraindividual telescoping, along with ontogenesis.

Brind's study constitutes an analysis of the development of
the readiness to be critical of certain more or less blatantly
erroneous sentences involving factual untruths and logical in-
correctness. Ten such sentences were randomly interspersed
with an equal number of correct statements and presented to
normal school children aged seven to fourteen, with forty subjects (boys and girls) at each year level.

For each of the sentences there was found an age level at
which (1) a kind of naïve credulity could be observed. Independent judgment seems to begin with (2) uncertainty or diffuse distress. Next comes a series of stages of criticism
proper, characterized by growing discrimination between the
correct and the incorrect sentences. The earliest explicit type
of criticism usually consists in (3) stereotyped rejection, or
global, en bloc negativism. Examples are "No, " "Bad, " "Can't
be. " This is followed by (4) specified rejection. By this is
meant a circumscribed statement of the weak point in the statement. Taking as an example the sentence, "When it rains, it

is dry, " a case of specified rejection would be "Not dry. " A still higher level is that of (5) positive correction, given for our example by such statements as, "It is wet. "

The most advanced stages transcend criticism in the narrow sense of the word. We shall subsume them under the term (6) higher-order criticism. Here we find a series of parallel subcategories. Perhaps the most important is (a) constructive interpretation, in which an attitude of making the best of the material presented is taken. For our sample sentence, an example is, "Yes, it is dry if there is a roof overhead. " Perhaps interpretation is the result of a kind of social consideration of the person who stands behind the statement in question. In this case the attitude toward the author is a benevolent one, reflecting a trust in the good senses of one's fellow men together with an implicit understanding of the intrinsic limitations of communication and the ensuing need for the use of indirection and metaphor. Other subvarieties of higher-order criticism are given by various types of (b) negative reference to the author of the sentence. An example is, "The man who has written this must be crazy. "

The various forms of criticism proper and of higher-order criticism just listed are followed by a last stage in which the subjects merely express (7) amusement, disinterest, or even surfeit and disgust in the task without taking the trouble of dealing with it in any specific way.

The sequence of the seven attitudes outlined in the preceding paragraphs was found to hold, first, when the total frequences of respective utterances of all subjects, grouped by age levels, were compared with one another. The lower the number in our list of critical attitudes, the lower the age of the onset, the peak, and the decline of the relative frequencies. For our type of material, higher order criticism does not occur before the age of nine and fails to come into its own before age eleven. (Although it is true that such a cross-sectional approach is not ontogenetic in the literal sense of the word, this type of approach is widely accepted as a more practicable substitute for the truly ontogenetic, longitudinal approach, at least for a good many types of child-psychological problems.)

The same sequence was found, second, when the responses of children of a similar age level to sentences of different de-

grees of difficulty were compared. This subsumes the material
under what we have labeled above task-induced developmental
hierarchy. And third, the same sequence of stages was found
when the responses given at the beginning of the experiment
compared with those given by the same subjects at later stages
of the same experiment. Protracted occupation of the same
kind thus acted in the way of an artificial evolution, in the manner characterized above as practice-induced developmental
telescoping.

In trying to relate Brind's results to the type of genetic series which we have come to know as the history of science and
of ideas, we may for the moment ignore the fact that the crucial
part of her material is false rather than true; as we all know,
most of the really decisive restructurings in science were for
long periods treated as falsehoods. We are also reminded of
Schopenhauer's saying that each truth enjoys but a short life
span of interested acknowledgment that lies between its being
ridiculed as absurd and its being discarded as trivial. Our
categories describing the sequence of critical attitudes are
flexible as to content, yet fairly detailed in elaborating the
various formal steps. We may therefore in this case confine
our comparison to relatively indirect, generalized, or casual
evidence.

The preliminary stage of naïve credulity may be seen as par-
alleled by the gullibility so frequently observed at the popular,
prescientific level. Among eminent scientists observant of
their own creative processes, Helmholtz--at a dinner in honor
of his seventieth birthday--has referred to the thoughts "creep-
ing quietly into my thinking without my suspecting their importance. " The ensuing sequence of astonished bewilderment,
emotional and total refusal, petty criticism, and the eventual
turning to positive and "higher" forms of criticism is, within
the field of psychology, best exemplified by the changing attitude toward psychoanalysis. After half a century of vacilla-
tion we have now reached the stage of interpreting psychoanalysis in a constructive manner and of trying to fit it into the
pattern of standard experimentation. [9]

The stage of constructive interpretation is that of real supe-
riority. The partial truth implicit in so many of the "creative
errors" in the history of science becomes clearly emphasized.

Prominent among these is the opening up of a new problem area by means of a false solution. [10] So far as the history of psychology is concerned[11] we may think of Mesmer's fantastic "animal magnetism" that turned into the sound doctrine of hypnotism, or of Gall's "phrenology" that gave rise to the serious treatment of brain localization of mental faculties, eventually helping to revise the classification of these faculties as well. In commenting on Brind's above-described results, Otto Selz, prominent member of the so-called Würzburg school of thought psychology, stressed the facilitation of problem solution by preceding mistakes in the history of science and of technology, along with the validity of the same principle for the dynamics of individual thinking processes. [12]

The Gestalt psychologist, Köhler, has stressed the importance of "good error" in the problem-solving behavior of chimpanzees, [13] thus furnishing a parallel from what Sander would call actualgenetic material at the phylogenetically lowered level of animal psychology. If a banana placed outside the cage could not be reached by using a single stick, some of the apes would spontaneously push out another stick by means of the first so that an optical bridge was formed connecting the hand with the banana. This was found to be a step preliminary to the crucial invention of mechanically joining two sticks of different diameter by inserting one into the other end by end, thus manufacturing a suitable tool for obtaining the fruit. The error of confusing the visual and the mechanical aspects of reality committed by the apes has a counterpart in children when they confuse meteorology and astronomy, and in the history of science when the firmament is conceived of as a dome with the stars moving along its surface.

A sequence recently observed in the development of the views regarding the role of "metaphysics" and of other imperfect instruments of knowledge, such as subjective introspection or speculation as they concern psychology, is also to the point here. Classical writers of the school of logical positivism have stressed the untestable, "meaningless" character of metaphysical statements, taking the latter in their literal meaning. One of the most totally rejective formulations comes from Philipp Frank[14] when he speaks of the metaphysics of yesterday as the common sense of today and the nonsense of tomorrow.

Such admonitions as "Positivism, not negativism!" perpetrated by one of the members of the next generation of the school, Herbert Feigl, [15] are symptomatic of a trend toward more advanced forms of response that move in the direction of "making the best" of any and all of the products of human imagination. As this writer has pointed out elsewhere in greater detail, [16] the so-called operational redefinitions of traditionally metaphysical or subjective concepts such as "purpose" (by Tolman), "hypothesis" (notably in animals; by Krech), "intentionality" (by this writer), and so forth, which since 1930 have been superseding John B. Watson's iconoclastic early behaviorism, must be seen as but another facet of the type of constructive interpretation which Brind has found to be an advanced form of criticism in her children. It stems from the conviction--appropriate to one engrossed in a discipline which still is in its adolescence--that not only science at large, but each science individually, and each unit of knowledge within it, must grow like an organism that passes through a variety of stages of increasing self-sufficiency and maturity; and that in this process the "genidentity" (to use a term of Kurt Lewin) of the problem line is maintained through the most bizarre and unrecognizable metamorphoses.

Theorizing in Children and in Early Science

Turning now from genetic sequences of such general attitudes as criticism to the more material content of science, the work of Piaget on the theorizing and the world view of children deserves particular attention. Since most of Piaget's books relating to the subject have been translated[17] we shall confine ourselves to a few examples. The validity of generalization of Piaget's findings regarding age levels or culture has been questioned as well as his adequacy of method; this writer feels, however, that so long as any children of any age anywhere produce the views found by Piaget with fair assurance of spontaneity in the sense of the absence of specific prior teachings, as seems to be the case, the major point has been maintained.

Most of Piaget's parallels relate to the history of physics. They are based on the free-flowing speech of, and interviews with, children, mostly under the age of ten. Simple questions wére asked, e. g. , "How does it come that a boat floats on the water?" In some cases the children had to predict the outcome of simple experiments, e. g., whether a stone or a match would sink or not. The results show a remarkable specificity, as well as in many cases a remarkable deviation from the views which could reasonably have been imparted to the children by adults.

One of the central concepts in the terminology of Piaget is "conservation." The stable world of the adult as well as his physics is based upon the discovery of a number of invariants. Piaget maintains that these constants and laws enter the scope of the child slowly and under great difficulties. To begin with, up to the age of about eleven or twelve, children do not know of the principle of inertia. In order to explain the flight of a projectile, children below age ten will in most cases assume that the projector produces air which is constantly acting upon the projectile. For example, a ten-year-old child thought that without air the projectile would fall to the ground immediately, and that "the air went all the time and pushed it along." In fact, it is the very weight (mass) of the projectile which by resisting the push of air is seen as ultimately stopping the movement.

Piaget points out that this explanation of continued movement is strikingly similar to the famous explanation suggested by Aristotle, known as the theory of antiperistasis. According to this theory the air plays the part of a motor. Shaken by the projectile as it issues from the sling of the catapult, the air flows after it and drives it along. After the contact has been lost the original impulse is transferred to the medium traversed by the projectile in a way similar to magnetism. This faculty decreases at a distance because of the resistance in the actual mass of the projectile, its natural weight. Aristotle thus reveals himself as being a long way off the principle of inertia as we have known it since the days of Galileo and Descartes, and as it is gradually transmitted to our children through cultural channels. For Aristotle as well as for the child the world is filled with spontaneous movement and living forces including those of the air, that is, with immanent animism and artificialism. Motion cannot occur without a moving agent permanently acting upon the thing in motion. Weight by itself, on the other hand, is seen as but an obstacle to movement.

Another variety of this type of explanation found both in Aristotle and in children is the theory of the reflux of air behind the projectile. This theory runs in a vicious circle. It is assumed that first the air in front of the projectile is put under pressure by the projectile; escaping from that pressure the air next flows behind and eventually goes on to push the projectile forward. Similar is the explanation given by children of the movement of the clouds. They are supposed to advance owing to the air that they themselves produce and that moves behind them and pushes them along.

An example from psychology is given by the resemblance, again to rather minute detail, between children's theories of perception and those of Empedocles or Plato. According to the latter, vision is due to light given out by an object as it meets the flashes of light emanating from the pupil of the eye. Similarly, children about six years of age tend to confuse "seeing, " which according to them occurs outside the eye, with "giving light. " Street lamps are able to see. Unless they are shut, as they are at night, eyes give light like lamps. This material by Piaget ties in with an observation by G. Stanley Hall concerning a child asking his father: "Why don't our looks mix when they meet?"

Concerning the development of the concept of weight Piaget distinguishes three stages, at least two of which are pre-Archimedean. During the first stage it is implied that specific weight is the same for all bodies, so that weight and volume mean pretty much the same thing. A small piece of wood is considered necessarily heavier than a pebble of slightly lesser volume. In the second stage bulky objects no longer are conceived of as necessarily heavy. Yet the concept of specific weight is not yet correctly envisaged, different objects merely being regarded as made of more or less condensed or rarefied materials. One of the children, five and one-half, states that the pebble is full and heavy; wood is much heavier than the equal volume of water because it is packed whereas water is liquid. Weight is here seen as dependent on condensation and solidity. Traces of similar views can be found in Anaximenes and Empedocles. Eventually, but not before the age of nine or ten, a clearer conception of specific weight can be found at least in the case of such common substances as water, wood, and stone, overruling the earlier emphasis on compactness. Prior to this final stage, the weight of water is brought into the picture primarily as the cause of a current that is assumed to keep such objects as boats floating. In fact, according to very young children, boats float not because they are light but because they are heavy, thereby giving rise to strong counterforces and movements in the water that keep the boat floating. A kind of struggle between the water and the immersed object is envisaged, leading to the prediction that stones will float whereas matches will sink. In a slightly more advanced stage, boats are assumed to float because they are lighter than the total mass of water in the lake. Not before the final stage are they correctly perceived as floating in consequence of the fact that they are lighter than water at equal volumes.

Furthermore, children show difficulty in establishing for themselves the principle of conservation of weight. Objects are assumed to lose weight as they grow in bulk, for instance from a pellet to a bowl. The first thing grasped is absolute weight, and much later follows relative weight; the sequence from absolute to relative is one of the characteristic features of mental development in general. In a way similar to the pre-Socratics, it is assumed that air, fire, smoke, steam, and water have the power of transforming themselves into one another. This assumed possibility of transformation of anything into anything else may be largely responsible for the appeal held by the world of the fairy tale in the esteem of children.

In essence, any nonrecognition of conservation serves to make abstract issues unduly concrete and leads to confusion. This becomes especially clear in considering the problem of conservation for such mathematical variables as number or volume. Max Wertheimer, the founder of the school of Gestalt psychology, has searched primitive cultures hardly touched by the dawn of science for evidence of the contamination of early concepts of number and has come up with an impressive array of instances of the use of different terms for the same number when the elements are arranged in different patterns or when different kinds of objects are involved. [18] Piaget and his collaborators [19] have observed, among other

things, that the scattering of a group of objects over a larger
area than that of the original distribution tends to induce children to claim that now there are more, even if the scattering
has been performed in plain view of the child. As to volume,
the pouring of a liquid from one container into two smaller ones
tends to give the impression of an increase, or else of a decrease, depending on whether there is cognitive assimilation
to the increase in the number or to the decrease in the size
of the new containers, as the more dominant feature of the
change.

With all this, we have moved into the twilight region between concepts and percepts. It is of crucial significance that
educated adults, while freeing themselves from illusory en-
tanglements of number and volume so far as their thinking and
scrutinized observation are concerned, can be shown still to
fall for them in their more direct perceptions. [20] The thinking
of children, and that of prescientific primitives, thus turns
out to be in important respects similar to that of intuitive perception, both adult and childlike. By intuitive perception we
mean the fairly autonomous primitive instrument of cognition
within ourselves as scientifically minded adults that functions
with speed and spontaneity. To apply the terminology of Werner, [21] "diffuse" cognition is superseded by "articulate" cognition; and this is the case in ontogenesis, in the phylogenesis
of science, and in a simultaneous genetic hierarchy within the
fully matured individual.

Egocentrism and Level of Abstraction in Concept Formation

To the scientifically trained mind the concepts under discussion in the preceding section, such as energy, weight, number,
or volume, will appear as paradigms of objective, almost pre-
ordained modes of thought, and any deviation from them will
likely appear as error or illusion. Yet there are other areas
in the realm of concept formation where greater latitude seems
to exist; their study brings out some of the more subtle points
of what we may call competing "viewpoints" in the establishment and application of concepts.

A convenient and effective method of studying this problem
is to present pairs of common words and to ask for a supra-

ordinate concept. A study of this kind was conducted as a doctoral dissertation by A. Spielmann-Singer at the University of Vienna under the direction of Charlotte Bühler and of this writer. [22] Special attention was paid to potential relevance for the emergence of the scientific point of view. Children from four to ten years of age (ten boys and ten girls for each age level) were presented with such word pairs as "streetcar and automobile, " "spinach and lettuce, " "cup and kettle, " and so forth. The children were asked to find a word which would "fit both of them. "

The results may be discussed under two major headings. One concerns the particular content of the guiding viewpoint of concept formation, with special emphasis on what Werner has described as "syncretism" and Piaget as "egocentrism, " that is, the intrusion of subjective aspects into the objective classification. The other concerns specificity versus generality, or what is frequently labeled level of abstraction.

As to subjectivism, the youngest of our children showed a predilection for (1) classification in accordance with pleasantness versus unpleasantness; for example, "streetcar and automobile" was frequently answered by such phrases as, "They are fun. " For our type of material, practically no answers of this type were found after the age of seven. Next was (2) concentration upon the practical usefulness of the objects in question, for example when "lettuce-spinach" was responded to by "to eat. " There follows further detachment from the ego in the sense that (3) communality of sphere of life, of milieu, or of other perceptual grouping is being stressed; an example for our last-named pair is, "They grow in the garden. " For our material, this type of response reaches its peak at the age of six. The most mature type of response concentrates upon (4) characteristics resting in the objects themselves. Here we approach most closely what is commonly called the objective or science-type point of view, or what in German is labeled sachlich.. Examples are the description of "lettuce and spinach" as "green, " or as "vegetables. "

It goes without saying that in many instances the fourth, or objective attitude may coincide with the second, or utilitarian, as for instance in the case of "cup and kettle. " Here we have to do with human artifacts made for particular purposes. Yet

even in such cases it was possible to distinguish different degrees of egocentricity and to classify the answers accordingly.
For our material, objective, sciencelike responses are rare
in the age brackets from four to six; they jump from 8 per cent
at the age of six to 52 per cent at the age of seven, however--
obviously in connection with entering school--and they reach
90 per cent at the age of ten.

Parallel hierarchies from subjectivistic to objectivistic types
of classifications are abundant in the history of science. The
popular prescientific classification of animals into "harmful"
and "beneficial, " or their classification in terms of common
appearance or milieu (whales grouped with the fish) or in terms
of other phenotypical characteristics come to mind at once.
The geocentric world view is more perceptual--or "pictural, "
as Philipp Frank[23] would say--and hence more egocentric or
subjective than the heliocentric. In fact, all the great "Copernican revolutions" in the history of science are manifestations of what Freud has called "retreating narcissism, "[24] Copernicus himself renounced the idea of the dominance of man's
domicile over the heavenly bodies; the second of these revolutions, by Darwin, dethroned man among the animals; the
third, by Kant, challenged the subject's cognitive mastery of
reality; and still another, by Freud, exposed the conscious
mind as not even being master within its own personality.

Let us now turn to the second aspect of the development of
concept formation, specificity versus generality, or level
of abstraction. In Spielmann-Singer's data the concrete type
of answer occupies a particularly basic position. Between 30
and 35 per cent of the children in the age groups from four
to six gave answers that were erroneous by being too specific.
An example is "made of iron" as a reply to "cup and kettle, "
leaving out all cups or kettles made of materials other than
iron (or metal). This type of answer does, however, fade out
almost completely after the age of seven so far as our material is concerned.

Overspecificity is superseded by another type of inappro-
priateness, overgeneralization, with a peak between 20 to 25
per cent at the age levels of seven and eight. An example of
such an answer is "object. " Since responses of this kind are
apt to embrace both of the concepts presented, overgenerali-

zation is not always an error in the strict sense of the word. But from the adult standpoint overgeneralization must be regarded as an easy escape with little value for the specific problem at hand. Overgeneralization is a transitory type of response. Within our material it plays very little role at first and shows a marked tendency to decline after age seven and eight.

From then on, there is increased emphasis on what the logician calls the *"genus proximum,"* that is, on concepts fully embracing the original items but avoiding as much as possible going beyond. For the pair "streetcar-automobile" such an answer is exemplified by "vehicle." At the age level of ten, 87 per cent of the answers fall in this category.

The sequence from the concrete to the very abstract and down again to the medium level of the *genus proximum,* which we have observed in the quasi-ontogenetic data just presented, can also be found in what we have introduced above as actualgenesis in the sense of Sander. Using the retrospective method as developed by the Würzburg school, Willwoll[25] studied this microdevelopment of individual problem solutions in educated adults. Pairs of words were presented and the experimental subjects were asked to find a supraordinate concept. The intermediate level of the *genus proximum* was found to be the latest and most difficult in most individual solutions except the more automatic ones. At the same time, finding this level is experienced as especially rewarding from the standpoint of the subject's self-imposed level of aspiration.

Both the ontogenetic and the actualgenetic findings just presented seem to be in conflict with the commonsense view of the development of scientific thought. According to this view the concrete level is the easiest to handle and the first to be acquired while each step up the ladder of abstraction involves added difficulty, so much so that the highest abstractions can be approached only by such princes of thought as the philosophers.

In reality, the vast generality of such universal "dichotomies" as "being versus becoming," or "mind versus matter," are probably relatively easy to come by. Indeed, they are an early product in the history of ideas, and there is evidence from child psychology that they are early in ontogenesis also.

The over-all picture is somewhat complicated by the fact that
such "metaphysical" concepts may have specific ties with the
world of concrete objects or events by virtue of their percep-
tionlike or "pictural"[26] (see above) or "metaphorical"[27] character, and that these ties in a certain sense undo the all-em-
bracingness of their meanings. But there can be little doubt
that science proper tends to move at intermediate levels of
abstraction. Much abstraction, but also much technological,
operational, or methodological specification is required for
the establishment of such concepts as correlation coefficient,
or momentum. Only by virtue of such specification may we
move from the metaphysical stage to the "positive" stage,
as Comte has envisaged it, or from the "Aristotelian" to the
"Galilean" modes of thought, as Lewin[28] has described them
with an eye on the particular problems facing psychology in
its historical development.

Perceptionlike Structures (Formalism) in Early Science

In the history of geography and of astronomy the theme of
"perfect" form has exerted recurrent and lasting influence.
Ancient geographers have shown their predilection for regularity of form by assuming that the shape of continents and islands was originally circular, triangular, or rectangular, existing irregularities being the result of deterioration in time.[29]
In assuming the circularity of the orbits of the planets, Copernicus was prompted by the same formalistic bias. Had
Tycho Brahe not shared this formalism, he would not have re-
futed the Copernican system as a whole on the basis of certain
inconsistencies with observation but would have revised it.
Not before another generation had passed did Kepler struggle
through to accept the ellipse in spite of the fact that it was
less simple.

It will be noted that, in terms of our above discussion of
stages of criticism, Tycho is an example of negativism,
Kepler, of interpretation of data. The final stage of criticism, ridicule, in the present case of form, becomes myth-
ology, best exemplified by Fechner's satire *Vergleichende
Anatomie der Engel* (1825). This presents the argument that

since the sphere is the most perfect form, the angels, as the most perfect beings, must be spherical.

There is at least one area in which simplicity, circularity, and other features subsumable under "good form" are a reality, however. This area is the perception of form. An incomplete circle presented in short exposure tends to be experienced as a complete circle. Similarly, the negative afterimage of a square with one blunted corner will frequently appear either as a perfect square or else will show symmetrical corners blunted, and so forth. Generally speaking, under a variety of conditions of reduced stimulus impact, perceptual shapes will tend toward geometrically outstanding forms. Technically this is known as the "law of *Prägnanz.* " Most likely it is the result of dynamic self-organization within what such leaders in recent physiological Gestalt psychology as Köhler have termed "brain fields."[30]

We have already had occasion to characterize perception as a relatively primitive function in the simultaneous intraorganismic developmental hierarchy that makes up the adult human personality. By virtue of its formalism, early scientific thinking thus once more appears as a counterpart to a primitive subsystem of mature man.

The parallel can be expanded still further. Sander has subsumed short exposure and other reductions of the stimulus impact under his concept of actualgenesis,[31] in the sense that under these conditions cognitive acts are cut off in the midst of their natural maturation process, even though in most cases this process may not require more than a few seconds, or perhaps only fractions of a second, for its completion. In the light of this challenging notion, good form is but an intermediate one of several stages perception is capable of reaching on its own. Early scientific thinking would then not only display features of a perception, as such, but also of perception handicapped in the fulfillment of its own developmental potential.

Conclusion: Developmental Parallels within the History of Science Itself

It takes but a simple reapplication of the various aspects of what we have labeled simultaneous intraorganismic genetic hierarchy to view science in any cross section of time as a conglomerate of efforts moving at different developmental levels, depending on the age of the various disciplines, the difficulty of the problem, the time elapsed since work on the problem has begun, and so forth.

Speculations of this kind have led this writer[32] to attempt for psychology what one of his reviewers, Gustav Bergmann, has felicitously labeled "structural history writing" in contrast with the "pragmatic" tracing of historical influences in terms of intellectual biography, which characterizes the customary actuarial historical narrative. Structural history writing presupposes the classification of problems and modes of approach in terms of genetic sequences. One of the ways in which this can be done is by a pattern analysis of the scope of the varying scientific edifices that have occupied the attention of psychologists.

In executing this plan it first appears that the scope of the traditional subjective--that is, speculative or introspectionistic--psychology has, within the last three centuries, expanded from considerations confined to the internal life (Descartes) to a sensationist-peripheralist emphasis (English empiricism and associationism) and eventually to a stage in which the external reference of the inner states becomes the major issue. (Both Brentano's act psychology and Lewin's topological-dynamic psychology are relatively clearcut examples of this third stage.)

Next it is discovered that, with a temporal lag that decreases from centuries to decades as we progress in time, there is a parallel development at a methodologically higher, more "objective" level, which goes through a similar sequence of basic pattern or scope. It begins with the physiological psychology encapsulated within the body, as the early and middle nineteenth century knew it, progresses to the classical behaviorism of Watson with its more pointed emphasis on the sensory and motor peripheries and their interrelationships (stimulus-response approach), and so far has culminated in the more complex, more "molar" forms of behaviorism and physiological psychology (e. g., Tolman, Lashley), in which the long-range, or distal, external correlates or referents of observed or hy-

pothesized central or brain processes are the major issue.
From the genetic point of view both "objectivity" (scientific
exactitude) and "molarity" (that is, adequate complexity of
scope) are subsumable under the concept of difficult or late
developmental level, so that an advancement in one is often
found to go with a standstill or even a regression in the other,
at least temporarily.[33] The net effect is that each historical
cross section simultaneously embraces a variety of developmental stages.

Structural analyses and comparisons of this kind are in the
nature of things much more precarious than concrete fact finding or process tracing. Fortunately they can at least in part
be supported by quantitative analyses of historical trends in
psychology, which have of late become the vogue in a discipline very much aware of the necessity to introduce objective
methods along its entire front. Most notable of these is a statistical trend documentation based on a content analysis of
leading American psychological periodicals undertaken by
Allport in collaboration with Bruner.[34] It covers the crucial
period in the growth of psychology from 1888 to 1938. Our
claim of spiral recurrence of comparable historical sequences
at increasingly higher levels of either difficulty or exactitude,
which we have summarized in the preceding paragraphs, can
be borne out by means of Allport and Bruner's charts dealing
with the changing emphasis on certain fields or topics. There
we find a number of U-shaped time distributions, which may be
interpreted as parts of bimodal distributions; the latter in turn
may be taken to point to a shifting in the "kind" of endeavor
within the category in question. For example, the change from
the purely internal or peripheralistic physiological psychology
of the nineteenth century to the brain-and-achievement type
of approach which we have reported as the favorite of today is
reflected by a marked slump of interest in physiological problems at the beginning of this century.

In our opinion the fusion, just illustrated, between the structural, or broadly interpretational, approach to the history of
science and an adequately categorized, rigorous but comprehensive method of objective trend documentation promises to
render fruitful the application of genetic principles to the analysis of the history of science.

J. B. STALLO
AND THE CRITIQUE
OF CLASSICAL PHYSICS

BY STILLMAN DRAKE

"He is the author of the profoundest and most original work in the philosophy of science that has appeared in this country-- a work which is on a par with anything that has been produced in Europe. . . . It is, in fine, safe to say not only that the influence of Stallo's work will be a permanent one, but that it will also steadily increase. "[1]

Such was the appraisal of Stallo's *Concepts and Theories of Modern Physics* in an obituary sketch of its author written by Thomas J. McCormack, translator of Ernst Mach's principal works into English and probably, in his period, the most competent American judge in matters related to criticism of scientific theories. McCormack's verdict in his first statement was correct, even if the prediction he ventured in the second was not. Stallo's profundity and originality cannot be questioned. That his influence remained small among scientists was not remarkable, since he had opened his book with the words: "The following pages are designed as a contribution, not to physics, nor, certainly, to metaphysics, but to the theory of cognition. "[2] Traditionally the theory of knowledge was a concern of philosophers and not of scientists.

Events of the present century no longer justify a continued neglect of Stallo, especially from the historical standpoint. The revolution in physics that commenced about the time of his death in 1900, originating in philosophical no less than in strictly physical considerations, has vindicated Stallo to a great extent. We are by now accustomed to the fact that episte-

mological postulates may be no less important in shaping a
new scientific concept or theory than the newly discovered data that are to be incorporated in it. Such a state of affairs
had already been clearly envisioned by Stallo a full quarter
century before it began to intrude itself into the work of professional physicists, and he had even been able to indicate
certain directions in which alterations would have to be made
in classical theories in the interests of further progress. In
his own day, much of Stallo's reasoning rested upon grounds
that lay outside the accepted boundaries of physical thought;
subsequently those boundaries have been widened to include the
epistemological considerations Stallo utilized. To those interested in the evolution of scientific thought, a brief account
of this almost forgotten man and his work may therefore be
welcome, despite his apparent failure to influence directly the
course of events in physics.

In the *Concepts,* Stallo undertook to reveal the degree to
which outmoded philosophical doctrines still permeated science
in spite of all previous efforts to root them out. Stallo was no
longer a professional philosopher or scientist during the years
in which the book was being thought out, written, and published.
Yet the reader of the work cannot fail to be struck by the sin-
gle-mindedness with which the author restricted himself to
the subject matter of physical science and by his entire familiarity with its literature (including even those journals that
ordinarily fall only under the attention of men active in the
profession) as well as with that of philosophy. In view of the
technical nature and scope of his book, the great number of
editions and translations through which it passed between 1882
and 1911 can hardly be accounted for except by its wide dis-
semination among scholars, however little they may have cited or credited it, during that crucial epoch in the history of
physics.

The first edition of the *Concepts* was published by D. Appleton and Sons in New York as the thirty-eighth volume of that
memorable set of well-made duodecimos that collectively
bore the name of "The International Scientific Series." The
title page bears the date 1882, though the copyright date is
1881 and the book is said to have made its first appearance
in November of that year.[3] Simultaneous publication in Eng-

land (London: Kegan Paul, Trench and Trübner) was the rule
for these volumes, and a French translation followed immediately under the title *La matière et la physique moderne* (Paris: Felix Alcan). H. A. Rattermann states that a German
translation was published at Leipzig simultaneously with the
American and English editions. But despite Rattermann's having had access to Stallo's own collection of books (until 1885
at any rate), he seems to be in error on this point. The first
German version was the translation prepared by Hans Kleinpeter with the encouragement of Ernst Mach and published by
Barth at Leipzig in 1901.[4] Mach's preface to this volume, and
Kleinpeter's article on Stallo in connection with its publication, made it evident that the man and his work had been previously unknown in Germany. Rattermann states further that
translations were made into Italian (Bologna), Spanish (Ma-
drid), and Russian (St. Petersburg). A second American edition appeared in 1884 and a third in 1891; the third English edition had appeared in the previous year. The fourth printing of
the French translation is dated 1905, and a second German
edition was published in 1911. Thus it appears that no less
than fifteen printings of the work were made, in six different
languages, within thirty years of its original appearance.

The American editions of 1884 and later are of particular
interest, for they contain a long introduction prepared by Stallo in answer to his American and English critics and reviewer-
ers of the first edition. At the same time he made a few minor
alterations in and additions to the text, but no substantially
revised or expanded edition was ever produced. Stallo's il-
luminating and often sprightly reply to his critics was included only in these American editions, though the textual alterations were carried over into the English republications and
thence into the German translation. In this regard, the French
title pages are misleading; what purports to be the *quatrième
édition* is merely a reprint of the original text, containing
neither the 1884 introduction nor the revisions made at that
time. The preface to the French translation was supplied by
C. Friedel; of this, Mach said (quite justly): "That the book
is properly esteemed by experts may well be questioned from
any indications. The French edition is even provided with a

preface to which one can scarcely attribute any other purpose than that of weakening the effect of the book. "[5]

Mach became acquainted with the book and its author only a short time before Stallo's death. He first learned of its existence through a citation by Bertrand Russell;[6] and, he says, "I naturally took a lively interest in the man whose scientific aims so closely approximated my own. No one in England could give me any information about him, Professor A. Schuster of Manchester merely giving me his surmise that he might be an American."[7] With the assistance of Paul Carus, Mach at length established contact with Stallo, who was then residing in Florence. The short-lived correspondence into which they entered was soon interrupted by a serious illness of Mach's and was hardly resumed when terminated by the death of Stallo. Mach's personal judgment of Stallo, from whom he had meanwhile received a brief biographical sketch and copies of his two other books, is set forth in his preface to Kleinpeter's translation of the *Concepts,* from which the following excerpts are taken:

From his own account of his life, Stallo may be considered essentially self-taught, allowing himself to be led in his scientific studies only by the writings of the great discoverers of ancient and modern times. Without the personal direction of a teacher he was forced to resolve his own doubts through quiet and continual reflection. Thus he achieved an originality and independence such as is puzzling to orthodox youngsters of the modern school of physics, and toward which they are rather hostile. . . .

Through his philosophical and historical studies, Stallo was placed in a position to recognize in the presently received views of physics traces and elements of the outlook of past times which modern physicists in general take to have been long since vanquished, and which in undisguised form they would hardly recognize as their own. . . .

Since Stallo examined modern physics under the influence of this viewpoint, he necessarily perceived the scholastic-metaphysical elements which pervade it throughout. After this recognition, the gradual complete emancipation of science from these traditional, often primitive and barbaric, modes of thought appears as merely a necessary consequence of the further development, strengthening, and critical clarification of physics. I cannot completely agree with Stallo on all points; thus, I cannot join with him in his sharp and all-inclusive opposition to so-called metageometrical researches. But in the battle to eliminate from science the latent metaphysical elements I agree with him completely, and his works offer to me a valuable and welcome complement to my own. As a more important point of agreement I may adduce especially the rejection of the mechanic-atomic theory--not as a means of assistance to physical discovery and representation, but as the general foundation of physics and as a world view. We share also the apprehension of physical concepts such as mass, force, etc., not as peculiar realities, but as mere relations; connections of certain elements with the appearances of other elements. Through the assumption of the relativity of all physical properties and definitions, including those of space and time, there necessarily follows finally an agreement in the rejection of all expressions about the universe as a whole. My writings are directed, as conditioned by my training, my capacities, and my calling, to those physicists who are not averse from the logical clarification and philosophical deepening of their science. Accordingly I search primarily for scientific confusions and irrelevancies in particulars, in order to go on from here to a more general viewpoint. Stallo, on the other hand, takes the opposite path. Starting from very general observations, he brings to bear upon physics the propositions thus discovered. Both paths lead to almost invariably agreeing insights. Here I can but repeat what I have already said elsewhere: It would have been very heartening and beneficial to me, when I commenced my critical works about the middle of the.1860's, to have known of the related exertions of such a comrade as Stallo.[8]

I

In point of time, Stallo's first researches leading to the *Concepts* very nearly coincided with Mach's. Although not published in book form until several years later, the *Concepts* must have been written in large part by 1873, for in October of that year, Stallo published the first of a series of four articles in the *Popular Science Monthly* under the title "Primary Concepts of Modern Science"; these articles embody much of the central theme of the *Concepts* in words that recur almost without alteration in the text of that book. That these ideas had begun to take shape in his mind long previously is shown by passages in his paper, "Materialismus," first published in 1855 and reprinted in 1893.[9] Furthermore, it appears that he was prepared to continue the four articles mentioned above, since the final words in the last of these (January, 1874) promise a further article dealing with the application of the principle of conservation of energy to the field of theoretical chemistry. The promised article did not appear, but the indicated subject matter was published some eight years later in the concluding chapter of the *Concepts*.

Mach's contrast between his own approach to the critique of physics and that of Stallo is certainly correct so far as it goes. There were, however, other differences both in their aims and in their outlooks, which bore essentially upon their methods and results. One such difference is instanced by the more marked radicalism in Stallo's application of the principle of relativity of physical data. Mach was never willing to accept what might be called, for want of a less paradoxical term, an absolute relativism; later this became quite apparent in his opposition to Einstein and more particularly to some of the latter's disciples. Stallo, had he lived, would undoubtedly have taken keen delight in at least the special theory of relativity and would have hailed it as the next essential step in physical theory. Mach, as a professional physicist, to some extent had a vested interest in the data of physics and thus was inclined to temper his criticism when he could not perceive any alternative means of preserving that body of knowledge intact. Stallo, on the other hand, had no stake in the fortunes of those data; this appears repeatedly from his suggestions as to possible modifications in the concept of mass, in the form of conservation principles, and the like. A striking illustration of this contrast between the men is to be found in their respective treatments of an argument that had been advanced in 1869 by Professor C. Neumann, as demonstrating the necessity of some absolute body of reference.

Mach says:

The most captivating reasons for the assumption of absolute motion were given thirty years ago by C. Neumann. If a heavenly body be conceived rotating about its axis and consequently subject to centrifugal forces and therefore oblate, nothing, so far as we can judge, can possibly be altered in its condition by the removal of all the remaining heavenly bodies. The body in question will continue to rotate and will continue to remain oblate. But if the motion be relative only, then the case of rotation will not be distinguishable from the state of rest. All the parts of the heavenly body are at rest with respect to one another, and the oblateness would necessarily disappear with the disappearance of the rest of the universe. I have two objections to make here. Nothing appears to me to be gained by making a meaningless assumption for the purpose of eliminating a contradiction. Secondly, the celebrated mathematician appears to me to have made here too free a use of intellectual experiment, the fruitfulness and value of which cannot be denied. When experimenting in thought, it is permissible to modify *unimportant* circumstances in order to bring out new features in a given case; but it is not to be antecedently assumed that the universe is without influence on the phenomenon in question. In fact the provoking paradoxes of Neumann only disappear with the elimination of absolute space. [10]

Stallo, several years previously, had treated the subject as follows:

The reasoning of Professor Neumann is irrefutable, if we concede the admissibility of his hypothesis of the destruction of all bodies in space but one. But the very principle of relativity forbids such a hypothesis. The annihilation of all bodies but one would not only destroy the *motion* of this one remaining body and bring it to rest, as Professor Neumann sees, but it would also destroy its very *existence* and bring it to naught, as he does not see. A body cannot survive the system of relations in which alone it has being; its *presence or position* in space is no more possible without reference to other bodies than its *change of position or presence* is possible without such a reference; and, as I have abundantly shown, all properties of a body are in their nature relations, and imply terms beyond the body itself. The case put by Professor Neumann is thus an attestation of the truth that the essential relativity of all physical reality implies the persistence both of force and of matter, so that his argument is a demonstration, not of the falsity, but of the truth of the principle of relativity. [11]

When Mach calls Neumann's hypothesis of annihilation a "meaningless assumption" he refers to no more than the fact that it is never realized in experience; here Mach is asserting his sensationalistic philosophical views, from which his belief in the relativity of physical data is derived and by which it is limited. "Elimination of a contradiction" by such means appears to Mach pointless because a hypothetical contradiction in the above sense is subordinate in his eyes to his principle of economy of thought in the construction of physical theories, which program demands only the avoidance of concrete contradictions. It is only upon such assumptions that he is en-

abled to assert next that the circumstances modified in Neumann's thought-experiment are unimportant and thus to reject
Neumann's assumption of the noninfluence of remote bodies.
(And in view of Mach's previously cited statement, that all
propositions about the universe as a whole are to be rejected,
his appeal here to the possible influence of the universe upon
the phenomenon in question is at best a desperate expedient.)
In effect Mach simply avoids, and does not refute, the whole
logical fabric of Neumann's argument.

Now Stallo also calls Neumann's hypothesis meaningless,
but without any qualification as to "what is to be gained" or
what is or what is not "important"; he takes his stand firmly
upon the principle of relativity of all physical properties, not
excluding the property of existence itself. Thus Stallo exposes
the futility from a logical standpoint of Neumann's position,
for Neumann has attempted to address the relativists with an
argument that must be and remain meaningless to them.

The disparity between Mach's approach and Stallo's, as illustrated in this single example, may perhaps be accounted
for in the following way: Mach was primarily concerned with
describing the occurrences, persistences, and reappearances
of metaphysical biases among scientists, and with accounting
for their gradual historical elimination in accordance with his
principle of economy of thought. Stallo, on the other hand, ad-
dressed himself to the task of revealing the origins of such
ideas in scientific contexts and of explaining how they got there
in the first place and how they might be avoided. In this sense
it is especially true, as Mach says, that the works of the two
men admirably complement one another; it is almost as if
in stocktaking late in the nineteenth century, Mach showed
the scientists where they had been and Stallo pointed out to
them where they were likely to be going.

"Although the founders of modern physics at the outset of
their labors were animated by a spirit of declared hostility to
the teachings of mediaeval scholasticism," says Stallo in the
introduction to the second edition of the *Concepts*," . . . when
they entered upon the theoretical discussion of the results of
their experiments and observations, they unconsciously proceeded upon the old assumptions of the very ontology which they
openly repudiated . . . founded upon the inveterate habit of

searching for 'essences' . . . before the relations of words
to thoughts and of thoughts to things were properly understood. "
His object, he declares, has been

. . . to consider current physical theories and the assumptions
which underlie them in the light of the modern theory of cognition--
a theory which has taken its rise in very recent times, and is founded
upon the investigation, by scientific methods analogous to those
employed in the physical sciences, of the laws governing the evolution of thought and speech. Among the important truths developed by
the sciences of comparative linguistics and psychology are such as
these: that the thoughts of men at any particular period are limited
and controlled by their forms of expression; viz., by language; that
the language spoken and "thought in" by a given generation is to a
certain extent a record of the intellectual activity of preceding generations, and thus embodies and serves to perpetuate its errors as well
as its truths; that this is the fact hinted at, if not accurately expressed, in the old observation that every system of speech involves
a distinct metaphysical system . . . that philosophers as well as ordinary men are subject to the thralldom of the intellectual prepasses-
sions embodied in their speech. [12]

Proceeding with this program, Stallo identified four metaphysical assumptions so intimately linked with traditional views
of the uses of language that he termed them "structural fal-
lacies of the intellect. " These are the fallacious assumptions
that: (1) every concept is the counterpart of a distinct objective
reality; (2) general concepts and their counterparts pre-exist
to the less general; (3) the genetic order of concepts is identical with the genetic order of things; and (4) things exist independently of and antecedently to their relations.

Examples of these fallacious assumptions in physical science
he found on all sides, but particularly in the declared intention
of scientists to reduce all natural phenomena to mass and motion while insisting upon the absolute disparity of the two and
upon their separate conservation. That mass remains the same
whether at rest or in motion, Stallo recognized as an unwarranted assumption; that the conservation of mass had any independent meaning apart from consideration of energy, he
saw to be equally so.

But his chief target of attack was the atomic theory of his
day. To attribute indestructibility to atoms in order to explain
it in gross matter struck him as particularly ridiculous: "A

phenomenon is not explained by being dwarfed. A fact is not transformed into a theory by being looked at through an inverted telescope."[13] "A valid hypothesis reduces the number of the uncomprehended elements by at least one," he remarks, citing Zoellner.[14] But "some of the uses made of the atomic hypothesis, both in physics and in chemistry . . . replace a single assumption by a number of arbitrary assumptions among which is the fact itself"--that is, the fact which was to be explained by means of that hypothesis.[15] The kinetic theory of gases and the wave theory of light are subjected to searching criticism in the *Concepts;* the disparities between the atoms demanded by physicists and those required by chemists are displayed; the variety of ethers provided on order for the various purposes they were to fulfill is ridiculed, and the contradiction between the accepted phenomena of gravitation and the avowed rejection of action-at-a-distance is exhibited. Many supposed attempts to reduce the complex to the simple are shown to have been merely reductions of the unknown to the familiar, which is a very different thing.

Stallo's relentless relativism broke down, however, at one point; he was unable to admit the relevancy of non-Euclidean geometries to physical questions. In two chapters devoted to this subject he argues, much as did Poincaré in his popular writings on this matter, that the mutual translatability of all geometries precludes the necessity of abandoning Euclidean geometry for any scientific purpose, and that the endowment of space with properties capable of rendering any portion of it distinguishable from any other amounts to depriving space of that very essence which renders it distinguishable from matter. Since Stallo's arguments on this point are inadequate in any case to accomplish his purpose in raising them, he cannot be entirely forgiven for his essential departure here from his own basic tenets. The speculations of W. K. Clifford upon possible uses of the new geometries in physical theory were familiar to Stallo, but he rejected them as imputing material properties to space. It is curious, in view of his critique as a whole, that Stallo should have failed to recognize the inacceptability of the only alternatives to granting this sort of relativity between space and matter--either the abandonment of the concept of space entirely, or the assignment to space of a separate and independent existence. The latter alternative is, in effect, actually accepted and defended by Stallo; thus the geometrical section of the *Concepts* is completely out of accord with modern physical views.

The concluding chapters of the *Concepts* are devoted to a critique of the then prevailing cosmogonies and cosmological speculations, and to a commentary upon attempts to reduce all chemical phenomena to explanation by the principle of conservation of energy--attempts which the author felt were certain to succeed eventually, with the appropriate recognition of the interrelationship of energy and mass.

Such, in brief, was the program of Stallo's *Concepts*. The book was not well received by the critics of his time; it did not obtain a highly favorable evaluation from a scholar in the field until after his death, when Kleinpeter (who at the time was preparing a German translation of the book) wrote an article entitled "J. B. Stallo als Erkenntniskritiker, " which appeared late in 1901. [16] Subsequently Stallo's work received attention from Emile Meyerson, who referred to it often in his *Identité et Réalité* (1908); more recently Stallo has been cited by Rudolf Carnap in *Der Raum* (1922) and in the *Physikalische Begriffs-bildung* (1926), and by Percy Bridgman in *The Logic of Modern Physics* (1927). Perhaps it is only in the light of the hard-won physical knowledge of the twentieth century that an approach like Stallo's to the domain of theory construction can be truly appreciated.

II

The author of the *Concepts,* John Bernard Stallo, was born March 16, 1823, in the town of Damme, parish of Sierhausen, in southern Oldenburg, Germany. His father was a schoolmaster in that town, as had been his grandfather; so far as he could trace his family tree, all his forebears on both his father's and his mother's side were country schoolteachers.

My grandfather [Stallo told Rattermann] was my first teacher. He was a venerable old Frisian (Stallo is not an Italian but a Frisian name, meaning "forester"), who to his dying day wore the three-cornered hat, knee-breeches, and buckle shoes [of his profession]. He undertook my upbringing despite his more than seventy years of age, and was not a little pleased when I had learned to read and to do all sorts of problems in arithmetic before the end of my fourth year. To him I am indebted for my education in the ancient languages and in English; but French I learned from my father behind my grandfather's back, for he hated "Frenchness" with all his soul.

Stallo's education was well advanced when he emerged from school at his first communion at the age of thirteen, and he was then entered in the teachers' college at Vechta, which he could attend free of charge. There he had the opportunity of studying in the neighboring Gymnasium as well and in two years was ready for admission to a university. But his father lacked the means for that step, and "there remained to me only the choice of following the family profession of teaching school, or emigrating to America. This lay close to my heart, since my father's brother, Franz Joseph Stallo, had opened the way for emigrants from Oldenburg by going there in the early 1830's." This uncle had fled to America after having antagonized the authorities in Germany with his liberal political views. After an ill-fated venture in founding a community of his own, he had settled in Cincinnati where he was a successful inventor and printer.

Stallo arrived in Cincinnati in the spring of 1839, armed with letters of recommendation to eminent Catholic clerics in that area. Once there, it became a grave problem what he should do, since he was too well educated to become a laborer and yet he knew no trade. On the strength of his recommendations, he was given a post of instruction in a parish school; having emigrated from Germany to avoid becoming a schoolteacher, he promptly became one in America. His principal assignment was to teach German, and he soon set to work on a book to which in his later years he was wont ironically to refer as "my most brilliant literary success"--an elementary language text called *A. B. C. Buchstabir und Lesebuch für die deutschen Schulen Amerikas*. Published anonymously in 1840, it went through numerous editions and was widely adopted by schools in all parts of this country.

About this time the Athenaeum High School in Cincinnati was converted by some Belgian and French Jesuits into St. Xavier's College, and Stallo's abilities having come to their attention, they put him, at age eighteen, in charge of instruction in Ger-

man. During the first semester he was asked to assist in Greek
and Latin as well as in mathematics, and students flocked to
his classes. By arrangement with the authorities he was allowed to carry on his studies while teaching. The college had
a fine library in physics and chemistry and a reasonably satisfactory laboratory for the time, and Stallo accordingly made
use of his spare time for the study of those sciences under the
Jesuit instructor. He managed so well that in the autumn of
1844 he was called as professor of physics, chemistry, and
mathematics to St. John's College in New York (now Ford-
ham University), where he remained until 1847.

Shortly after resigning this post he published his first serious work under the title *General Principles of the Philosophy
of Nature* (Boston, 1848). The book had some merit in acquaint-
ing American scholars with the philosophical systems of several Germans, notably Oken and Hegel, but despite Rattermann's attempts to find passages of scientific importance in
this work (which he adduces as anticipations of Darwin's evolution theory), it was best described by Stallo himself in the
preface to the *Concepts:*

> I deem it important to have it understood at the outset that this trea-
> tise is in no sense a further exposition of the doctrine of a book . . .
> which I published more than a third of a century ago. That book was
> written while I was under the spell of Hegel's ontological reveries--
> at a time when I was barely of age and still seriously affected with
> the metaphysical malady which seems to be one of the unavoidable
> disorders of intellectual infancy. The labor expended in writing it
> was not, perhaps, wholly wasted, and there are things in it of which
> I am not ashamed, even at this day; but I sincerely regret its publi-
> cation, which is in some degree atoned for, I hope, by the contents
> of the present volume. [17]

Rattermann once asked Stallo whether the failure of the earlier book to gain acceptance had been his reason for forsaking
the teaching profession, and says that Stallo replied:

> See here, my friend, I had rather not discuss that. I found out that
> the American spirit was not yet ready for philosophy. Only its su-
> perficial growths flourish here; those which have deep roots and bear
> fruit cannot yet be so much as planted, for they will not grow to
> ripeness. I desired primarily to make sure of a secure living for
> the future, so I came back to Cincinnati. I wanted to become prac-
> tical, as the Americans are.

Upon his return to Cincinnati, Stallo hesitated between law and medicine, deciding upon the former on the advice of a fellow German who had found the latter neither a lucrative nor a pleasant calling among his compatriots in that city. Admitted to the bar in 1849, Stallo commenced a practice which was to continue uninterruptedly with the exception of a brief term on the bench (1853-55), until his departure from America in 1885. His most famous court case was the erudite and skillful defense of the Cincinnati public school board for dropping from the curriculum both Bible reading and hymn singing. Such practices had been offensive to Catholics, German Evangelists, Jews, and agnostics alike, since they had had to bear the burden of taxes for schools to which they could not conscientiously send their children, and which they were accordingly duplicating with various parochial schools. Stallo lost the case before the Cincinnati Superior Court (and "only the theologico-political Sanhedrin of Cincinnati" could so have acted, as he told his fellow townsmen), but was later vindicated by the Ohio Supreme Court. Stallo served for seventeen years as an examiner of teachers for the public schools, and during a part of this time served on the Board of the University of Cincinnati.

A great admirer of Jefferson, Stallo was a staunch Democrat in his earlier years, but as the slavery question became paramount, he forsook that affiliation and became one of the founders of the Republican party and campaigned for Fremont. At the outbreak of the Civil War he called upon the Germans of Cincinnati to form a regiment of their own, the Ohio Ninth, which was one of the first in the field and became known as "Stallo's Turnier Regiment." After the war he became increasingly dissatisfied with the selfishness and corruption of the Republican machine and in 1872 was a leader in the ill-fated Liberal Republican party, which attempted to nominate Charles Francis Adams. The disastrous "politicking" of Carl Schurz, his houseguest during the Cincinnati convention, resulted in the ridiculous nomination of Horace Greeley and utterly disgusted Stallo. In 1876 he rejoined the Democratic party and assisted the Tilden campaign; in 1880 he denounced the Republican party for having outlived its usefulness and becoming a mere machine for the furtherance of the ends of industrial politicians,

monopolists, and speculators. In 1884 he supported Cleveland
in the hot battle against high tariffs, and following the Democratic victory of that year, he was rewarded with the post of
minister to Italy.

Four years later the Republicans regained power and Stallo's assignment at Rome was abruptly ended, but he did not
return to the United States. Instead, he removed to Florence,
where in accordance with a long cherished dream he settled
down to the exclusive pursuit of his scientific and cultural interests. William J. Youmans, then editor of the *Popular Science Monthly*, seems to have expected this to result in a sup-
plemental volume to the *Concepts*,[18] but Stallo's only further
publication was a collection of earlier periodical articles,
addresses, and letters, which appeared under the title *Reden,
Abhandlungen und Briefe* (New York, 1893). This volume, compiled at the request of friends, gives us a fairly complete picture of Stallo's political views and contains in addition such
characteristic pieces as biographies of Jefferson and von Humboldt, short popular articles on science and philosophy, a
commentary on the English language, and an account of the
case against Bible reading in the public schools. The book was
highly praised by Ernst Mach both for its style and for its
content. Stallo died at Florence on January 6, 1900.

Stallo's character was marked by the depth of his convic-
tions, his love of knowledge, and his passion for freedom;
he had the reputation of being somewhat brusque and was not
lacking in personal courage. During 1862, when there was
widespread hostility toward Wendell Phillips, Stallo was asked
to preside at a meeting where Phillips was to speak. He was
most unsympathetic with Phillips' disunionist views and at
first declined, but when he heard that others whom he had rec-
ommended as more suitable for the chair had refused either
from pressure or from fear of the crowd, he consented to preside. The meeting had no sooner begun than a shower of mis-
siles greeted both Phillips and Stallo, who nevertheless stood
beside him throughout the ensuing turmoil.

Stallo's interests in music and literature were broad and
deep, and his home was an established calling place for distinguished visitors to Cincinnati. He had built up, in addition
to his law library, a collection of some five thousand volumes

relating to literature and science. His chief treasure was a single volume in which were bound copies of Kepler's *Astronomia nova, Stereometria, Harmonices mundi,* and *Tabulae Rudolphinae;* these were heavily annotated in the margins by a hand which was, to the best of Stallo's knowledge and belief, that of Kepler himself.

Teacher, lawyer, judge, and diplomat--such was the man whom Kleinpeter astutely declared to have "made a juridical approach to physics, first establishing the crime and then prosecuting the offenders. . . . Mach, Clifford and Pearson are indeed in the first rank of epistemologists, but . . . Stallo is the only one who has rigorously and systematically pursued a critique of knowledge in mathematics and the sciences."[19]

Perhaps the only other nineteenth-century scientific amateur of equal theoretical insight was the British jurist to whom E. L. Youmans referred when he first introduced Stallo to the readers of a scientific journal in America: "It has long been the honor and boast of the British bar that Mr. Justice Grove, the author of *The Correlation of Forces,* belonged to it; it is equally to the credit of the legal profession in this country that a member of it has cultivated scientific philosophy to such excellent purpose."[20]

CHARLES FULLER BAKER, ENTOMOLOGIST, BOTANIST, TEACHER, 1872-1927

BY E. O. ESSIG

Charles Fuller Baker, distinguished as an entomologist, botanist, collector, agronomist, teacher, and university administrator, was born on March 22, 1872, in Lansing, Michigan, as the second son of Joseph Stannard Baker, a major of the United States Army, and his wife, Alice Potter Baker. The family had a total of ten children of whom two were girls. Of the eight boys, two in addition to Charles became well known: Ray Stannard Baker as an author under the *nom de plume* of David Grayson, and Hugh Potter Baker as a forester, educator, and college president.

At the Michigan State College of Agriculture, from which he graduated in 1892, Baker came under the influence and instruction of Professor Albert John Cook, who had been born and raised on a farm in Owosso, Michigan, and had graduated from the institution just mentioned in 1862. Professor Cook was a most enthusiastic and competent teacher and one of the founders of American entomology as a profession. To his dynamic personality Baker responded in a remarkable manner. Cook once told me that Baker, as a student, spent nearly all of his cash for insect boxes, much to the embarrassment of his father, and that by the time he graduated he had several hundred boxes of specimens--a larger and more representative collection than was possessed by the college at that time. As an undergraduate he assisted Professor Cook in 1891-92 in the varied duties of research and laboratory assistant. Cook's enthusiasm and kindliness exerted a profound influence upon

Baker just as they did upon many others who studied under this great teacher years afterward. Cook's interest in Baker never ceased and he was responsible for many of the latter's early accomplishments.

Following graduation and upon Cook's recommendation, Baker received an appointment as laboratory assistant to Professor C. P. Gillette, a graduate of Michigan in 1884, and then head of the Department of Entomology, Zoology, and Physiology of the Colorado Agricultural and Mechanical College at Fort Collins. In this relatively unexplored Rocky Mountain region, Baker made very extensive botanical and entomological collections and began publishing his findings in these fields. In 1893 he was placed in charge of the Colorado forestry and zoology exhibit at the Columbian Exposition in Chicago.

Baker's first important entomological paper in coauthorship with Professor Gillette was, "A preliminary list of the Hemiptera of Colorado" (1895).[1] It was in this paper that he first listed the sugarbeet leafhopper collected at Grand Junction, Colorado, August 26, 1894, as *Thamnotettix (Jassus) tenellus* Uhler.[2] It became one of the most important injurious insects in the Rocky Mountain and Pacific Coastal regions because it proved to be a vector of the virus disease of sugar beets, commonly known as "curly top," which is one of the most serious and destructive diseases of this important agricultural crop west of the Rocky Mountains.

The years 1897 to 1899 Baker spent partly in Alabama, as zoologist in the Alabama Polytechnic Institute, and entomologist in the Agricultural Experiment Station at Auburn. At this time he investigated the San Jose scale and wrote two bulletins on it; he also wrote bulletins on the fruit-tree bark beetle and the peach-tree borer. He furthermore took part in the Alabama Biological Survey. During the second half of this period, i.e., 1898-99, Baker was botanist on the Herbert Huntington Smith exploring expedition to the Santa Marta Mountains of Colombia. This was a great experience for him, and an important factor in his decision of 1912 to return to the tropics--a step which undoubtedly resulted in his amassing great collections, but perhaps also hastened the end of his brilliant career. In 1899, following the completion of the H. H. Smith expedition, he is reported to have presented to the

United States National Museum a collection of American Hemiptera and Hymenoptera amounting to sixty thousand specimens. [3]

From 1899 to 1901 Baker was head teacher of biology at the Central High School in St. Louis, Missouri. In the following year, 1902-3, he studied with Professor Vernon L. Kellogg at Stanford University, where he obtained the degree of Master of Science in 1903. Kellogg at this time was busily engaged in teaching, in doing research on the systematic, anatomical, biological, and histological aspects of insects, and in preparing the manuscript of his *American Insects* which was printed in 1905.

Through the efforts of Professor A. J. Cook, Baker in 1903 was induced to accept the position of assistant professor of biology at Pomona College, Claremont, California. Since there were few students in entomology at that time, Baker began an intensive study of Western insects, and as a means of publishing his discoveries, he began the serial *Invertebrata Pacifica,* [4] which he financed out of his meager earnings and which entailed a tremendous amount of field and laboratory work. During this period he also published two of his most important papers, "A Revision of American Siphonaptera, "[5] and "The Classification of the American Siphonaptera. "[6] While at Pomona he received a very flattering offer to become chief of the Department of Botany at the new Cuban Experiment Station (Estacion Central Agronomica), Santiago de las Vaga, Cuba.

Franklin Sumner Earle, who had been charged with the establishment of this new station in 1904, stated in 1928:

Baker was the first man that I thought of in connection with the new staff. I offered, and he accepted, the position of chief of the botanical department, for I realized clearly the great importance of the little known field of tropical economic botany and felt that Baker was the best man in sight to take hold and develop it. He began with his usual vigor, established contacts with Botanical Gardens in all parts of the tropics, and soon had the beginnings of a great living collection of tropical economic plants. Many of his plantings still exist at Santiago de las Vegas [in 1928]. It was here in Cuba that he first became interested in the general problems of tropical agriculture, an interest that was to be of great importance for his future work in the Philippines. [7]

Baker told me that while he was in Cuba a large American tobacco company offered him an unusually high salary if he would undertake a program of plant breeding to improve the strains of tobacco. Because Baker believed that tobacco was not good for young people, and since he never smoked or otherwise used tobacco, he refused to accept the position.

While in Cuba Baker discovered the qualities and abilities of a young native boy named Julian Valdez, who must then have been about sixteen years of age. He took a liking to this lad and befriended and unofficially adopted him. The boy was bright, alert, and had a rare native ability as a naturalist. He was carefully trained by Baker and devoted all of his time to collecting. He had a photographic memory and needed only to be shown an insect or plant once to be able to return to any locality and procure duplicates. With experience his knowledge grew, and it was he who scoured the valleys and the mountains, the grasslands and the tropical jungles, in search of new things that kept Baker pinning, pressing, and labeling far into the night and early morning. When Baker returned to Pomona College and, later still, went to the Philippines, young Valdez accompanied him.

At Pomona, where I became acquainted with him, Valdez seemed quite out of place in the college surroundings and spent most of the daylight hours searching the deserts and the mountains for specimens. He was probably one of the most remarkable entomological and botanical field collectors of all time, and shares with his master to some extent the credit for the quantity and quality of the great Baker collections. However, I have been told that in the Philippines he lost much of his early competence and fell victim to various undesirable addictions of our civilization so that he became a great worry and a financial burden to his benefactor and friend.

From Cuba, Baker, in 1907, was drawn away to assume the position of curator of the botanic gardens and herbarium of the Museum Goeldi at Para, Brazil. In 1908 he also became director-elect of the Campo de Cultura Experimental Paraense. At Para with the aid of Valdez he amassed very large tropical, botanical, and entomological collections. Concerning his stay in Para, Baker told me that yellow fever was rampant; that European visitors, entertainers, and officials "died like flies";

that he considered himself lucky to have been able to resist
or otherwise escape the terrible malady; and that he was very
glad to leave. In spite of the danger he came away with a large
collection of tropical plants consisting of several thousand
herbarium sheets and some fourteen thousand specimens of
insects that were subsequently presented to Pomona College
in 1908.

At the request of Professor Cook, Baker again came to
Pomona College in 1908. It was at the beginning of my junior
year that I came under his singular guidance. Baker was a
handsome man of medium height, slightly built, and very active and graceful. His personality was dynamic and he had un-
bounded energy and alertness. I never heard him say a harsh
or unkind word to anyone, and I never left him without a height-
ened feeling of strength and enthusiasm. For his students and
associates he set a pattern of friendliness, honesty, frank-
ness, and industry unexcelled. He fashioned and attained all
his objectives with confidence and optimism, and these qualities made him a great and inspiring teacher. He gave me as
a research problem a study of the hemipterous family Aphidae,
remarking that "it was an important economic group of insects that needed more investigating." I accepted this suggestion and have continued to this day the unraveling of the complicated nomenclature and biology of this important and destructive family.

With his enthusiasm and confidence, and the untiring and
unselfish cooperation of Professor Cook, Baker accomplished
remarkable work at Pomona College. Many things including
room space and desk facilities, supplies, and finances were
needed to supply him and his students, who came to him as
if charged by a magnet. Funds for these were secured by Professor Cook directly from the college as well as from private
individuals and from fruit growers' organizations. Systematic
and life histories of insects were fundamental. For the citrus
fruit growers of the region, a new system of orchard inspection
was organized to conduct a survey of citrus insect pests, which
gave not only excellent field experience and remunerative employment to advanced students, but also surprising and rich
returns to the growers. As a result, during the four years
of Baker's second Pomona period, he and Cook attracted,

trained, and graduated a group of biologists of a quality rarely produced by any similar institution in so short a time.

Early in 1909 Baker explained to Cook the great need of serial publications, not only as an outlet for the work of students, specialists, and faculty, but also for the benefit of all interested in the biological sciences and especially in entomology and botany. Cook at once agreed and undertook, single-handed, to raise a sufficient amount of money by private subscription, to finance both a *Journal of Entomology* and a *Journal of Economic Botany*. The first number of the former appeared in March, 1909, and of the latter, in February, 1911. Another notable contribution was the organization of the Pomona College Laguna Marine Laboratory in 1912.

In October, 1911, the appointment of Cook as state horticultural commissioner of California with headquarters at the state capital, Sacramento, broke the cycle of the Cook-Baker entomological activities at Pomona College. With Cook four hundred miles away there was no one left to secure the cooperation and financial support needed to continue the many activities set in motion by these two great men. Even before the college authorities could realize this situation, much less take steps to relieve it, Baker, early in 1912, was offered the important position of professor of tropical agronomy in the College of Agriculture of the University of the Philippines, at Los Banos, by his good friend Dean E. B. Copeland. This offer Baker accepted, and he removed again to the tropics. It proved to be the last chapter of his life. Concerning Baker's leaving Pomona College, President James A. Blaisdell wrote (1927): "Across all the years Dr. Baker is remembered here as a man of the most remarkable power and influence. He left an impression upon our students that will never be effaced and many of them look upon him as the most stirring influence for good things that ever came into their lives."

It is doubtful that Baker took a single insect or pressed plant with him to the Philippines. Once again he was beginning a new phase. As a teacher he took advantage of the new situation, and his work with the students earned for him their confidence and devotion, so that also in this tropical region he trained a group of men who will do much to mold the future work in entomology and botany.

During the first World War he took a leave of absence for six months to act vicariously as assistant director of the botanical gardens in Singapore. Here he not only assisted in taking care of and studying the famous botanical collections but also became acquainted with the native flora of the Straits settlements.

In 1918 he was appointed professor of tropical agriculture and, in addition, acting dean of the College of Agriculture of the University of the Philippines and director of the Agricultural Experiment Station. In 1919 he became dean.

During his long residence in the Philippines he rendered advisory and consultatory services in the following capacities: associate editor, *Philippine Journal of Science;* associate editor, *Philippine Agriculturist;* associate editor, *Philippine Agricultural Review;* cooperator, United States Department of Agriculture; special agent, Bureau of Non-Christian Tribes; technical assistant to the director of the Bureau of Agriculture; and tobacco inspector, Bureau of Internal Revenue.

His botanical and entomological surveys embraced all of the larger islands of the Philippines, Straits settlements, and northern Borneo. Throughout his long stay in the Philippines he left only once to serve the botanical gardens at Singapore as already noted. Every day and most of the night was spent in his work. His entomological collections he kept intact, and they received by far the greater part of his spare time. Botany came in for a share also, and fungi in particular were collected extensively throughout the archipelago.

Dean Baker always lived simply and for many years before his death he lived in a Swali house in which were his personal laboratory and storage rooms for his collections. For the last eight years (1919-27) he gave half of his salary to fellow scientists in Europe who had been reduced to poverty as a result of the war.[8] Charles W. Hamilton, pastor of the Student Union Church in Los Banos, wrote: "One of the outstanding and beautiful traits of his character was his generosity. Many a poor student can testify to his timely assistance in some critical hour."[9]

Baker's powers as a teacher were on all counts phenomenal. According to Hoffmann (1927) he "often took students who were thought incapable of learning, and made of them worthy schol-

Charles Fuller Baker in his office at the University of the Philippines

ars. "[10] But in regard to his total achievement perhaps no one can speak with greater authority than his friend Dean Edwin Bingham Copeland, who was instrumental in bringing Baker to the Philippines and the tropics again, and whom Baker succeeded as director of the agricultural activities of the College of Agriculture. His statement follows:

> Carl Baker ran a great race. The pace never flagged. He took no time out, never seemed to have to rest. He planned vacations; resigned his office in anticipation of necessary rest; but always that was a little way ahead, a tomorrow. . . .
>
> He was a naturalist; incidentally and rather as a diversion, a scientist. Scientists have the stage today, and naturalists have become few. . . .
>
> He was a teacher. Such zeal as his had to be contagious. Such vistas as he opened up demanded, rather than invited, exploration.[11]

The true story of Baker's life at Los Banos will probably never be fully related; however, it is only fair to his many friends in the Philippines to say that they have accorded him praise unstintingly. Colin G. Welles has given us this graphic picture as to how he actually lived:

> Five or six years ago, when I knew him as well as most men ever came to know him, Baker was living in a bamboo "bahai" on the

outskirts of the dank little village of Los Banos, forty miles south
of Manila. There, in his two rooms among the tops of palm trees,
with the stench of his neighbors' pigs and caraboos floating up through
the cracks in his floor, he made additions to his superb collections
of insects and fungi, and "thanked the Lord daily" for the ships which
brought him letters by the scores of unseen, unknown friends who
had come to know and revere his solitary work as a scientist.

Though he was then only a little over fifty years, fever and a hundred tropic diseases had wasted his body and parched his skin, so
that he looked more than seventy--very white of hair and intense
of eye.

Baker lived apart from the faculty of the College of Agriculture
of which he was dean. Between him and most of us was an intangible
though not unfriendly something which kept him from knowing the
men intimately. [12]

The general tone of Baker's life in the Philippines together
with his aspirations and disappointments are vividly depicted
in the many letters he wrote me during the period following
my graduation from Pomona College in 1909. The following
extracts are revealing:

If you are finding yourself buried up to the ears in work it should
make you very happy [Claremont, Calif., September 20, 1909].

Count that week a poor week toward your ultimate success that
does not interest some new man in your work or raise some new
champion for your cause [Claremont, January 7, 1910].

The many years bring ripe experience and a somewhat broader
view, but finally energies relax, and the keeness of youth becomes
blunted [Claremont, March 11, 1911].

The opportunities in the tropics are so colossal that I am astounded
more and more. If only living was better, and *good support assured*
--it would be the greatest work on earth [Los Banos, Christmas,
1912].

I have kept Julian in the field pretty constantly with tremendous
results. About a half dozen of the best European mycologists are
constantly busy studying my fungi! About forty[13] of the most active
entomologists of America and Europe are busy on my insects with
tremendous results.

The matter already published would make several good sized books

and there is more yet to follow. I have found time to get out a number
of papers myself. Though I am more successful as always in getting
others to turn out really great pieces of work than I am in turning
them out myself. Though I would if I could possible squeeze any more
activities into my life. I am hoping in my old age to really settle
down and develop some of the work I have in progress.

I am going to try to finish out four years ending in August (1916).
But my health is very poor and I have put in my resignation. I have
no plans for the future and I don't want any for awhile. I must have
rest and change! I do not expect to return to California--in fact,
can't.

I tell you, Essig, I never was and never will be a "professor" in
the classical sense but just a student among students willing to introduce them freely to my own interests and willing to help them
and stand by them to the limit. Me too--I hate the ordinary dicta-
torial type of pedant who holds himself above and aloof from his
students. He doesn't *get* them and he doesn't lead them on to great
things. Thank heaven my students have first of all been my *friends*.
I wouldn't have had it any other way for the world [Los Banos, February 9, 1916].

I have continued [through the First World War] correspondence
and sendings to Europe--but everything goes horribly slow--almost
wears out one's patience. Some of my old-time friends and collaborators in Europe have died since the war broke out--and with conditions as they are it is difficult to get some of the old work into new
hands. Kerremans of Belgium is dead, Dr. Rehm of Munich recently passed away, and just now I get the news of the death of Dr.
Poppius, who was far and away the best living authority on the Miri-
dae. This latter work will, most fortunately, be taken over by Dr.
Bergroth--who already had a great load of my work on his hands.
I now have about fifty entomologists hard at work on my Philippine
material. I think this comes pretty near "going some." Such a gait
is calculated to produce real results all along the line--in the space
of one man's lifetime. I have not the slightest idea how long I shall
be here, nor what the future may hold for me. Everything is in a
very disturbed and uncertain state, as it has been for several years.
Passing the islands over to the Filipinos, puts most of the Americans here "in the air." It seems that the robe of the Wandering Jew
must have fallen on me, and that my fate is to wander--wander far
and wide. In fact, the farther and wider it is--the better it suits me.
. . . In the meantime we are situated as you and everybody else is
--with only one duty, to work like "seven of a kind" while minutes
pass--make hay while the sun shines.

The rapid replacement of Americans by Filipinos in the University created a situation that caused Baker to cast about for a place to take his insect collection and to spend his retiring days working upon it. His first thought was to dispose of the collection to the California Academy of Sciences at a reasonable sum together with a suitable salary that would keep him comfortable the rest of his life. He needed to go to a cooler climate in order to recuperate and to regain much of the mental vigor and physical health that he had sacrificed in the tropics. He valued the collection at $50,000, which he offered in exchange for a nominal reward and modest yearly salary during the remaining days of his life.

I could not present them to the Academy outright because all my savings were buried in them. On the other hand I stated that if the Academy would not pay for the collection as such I would still be glad to come and turn over the collections on the basis of a much larger salary than would otherwise be considered, accepting this as covering the transfer of collections and distributing the consideration over a series of years which would be very favorable to the Academy.

A large collection of this kind could only be moved with any degree of safety from here between January 1st and the last of April. The other months being too humid [Los Banos, June 6, 1926].

Early in July, 1926, Baker received an invitation to attend the third Pacific Science Congress to be held in Tokyo, Japan. The Philippine government declined to pay his expenses as a delegate. After some attempts on the part of his friends to raise the money and a great deal of correspondence and discussion, the National Research Council agreed to pay his transportation. However, this arrangement did not suit Baker and his response to the offer was as follows:

I deeply regret the way matters stand--as I always regret unfavorable conditions in all cases. There is no way that I can see now in which these unfavorable conditions may be modified. Even had matters remained in the same shape as they were in at the time of your first kind invitation, I should probably have been forced to regretfully decline even the assistance of transportation. I have been here in a hot climate 14 years *continuously* and no longer have even the necessary clothing or travelling equipment to make a trip to Japan; and the personal necessary expenses of several hundred pesos for this purpose, I would be entirely unable to incur this year, excepting in the one event of complete liquidation and permanently quitting the work here.

Since you wrote, still other new considerations have arisen. As time has gone by and the political situation has become more and more strained, the situation of the few Americans in the service who are working *entirely under Filipino control* has become more and more precarious. The proposed Bacon and Kliess bills seem to have greatly increased the anti-American feeling--at any rate we find our position here far more difficult. It has progressed to such a point in my own case that the entire probability is that I shall submit my resignation at a very early date. The Board of Regents is largely Filipino and I understand that a number of them have expressed their desire for a Filipino dean in my place. Everything being so very unpropitious and the end so very near, I should now be unable to make the trip to Tokyo in any event unless it was a matter of leaving the Islands entirely and this can hardly be consummated by October even if I should begin to pack immediately. My collections and scientific materials *fill a house* and the labor of preparing them for shipment and getting them safely started will be *stupendous!* Where they can go to find a resting place in such a way that my association with them will continue I do not know. The great organization of specialists built up about the work and the great efficiency and productivity with which it has been operating, will make its enforced suspension little short of a tragedy!

I made the most strenuous efforts to find a place for permanent headquarters and it seemed for a time that that might be secured at the Academy of Sciences in San Francisco, but negotiations in progress with the Academy halted for reasons unknown to me, and I am therefore very much in doubt as to what the future may bring forth [July 12, 1926].

At the University of the Philippines the two surviving Americans found things becoming more difficult, and it was now only a matter of a short time before they would be compelled to seek employment elsewhere.

I rather dreaded to fight through for another season. It means tightly closed rooms and usually charcoal fires going day and night. Where this is not done collections are soon practically ruined by mould as has happened with the Bureau of Science collection in Manila. In mine the mould doesn't have a chance to get ahead. The only mould to get in comes on exchange specimens from other collections (of which I have great numbers) and these get special treatment without delay.

There are now only two Americans left here at Los Banos in technical work--myself and Dr. [R. L.] Pendleton in soil Tech.--an old Berkeley man. Both of us would be glad to "move on." Conditions constantly grow more difficult for us as you might imagine from what you see in the papers. I may feel compelled to resign by the end of this year anyway, and then I may follow Horace Greeley's advice, "Go west, young man, and grow up with the country!" I will have to get a change before long or become a candidate for "Salt Creek!"

While we wait work is pushed forward even more rapidly than usual and we stir up things locally even more violently than usual-- all of which indicates that at least our pep doesn't fail [August 12, 1926]!

The last letter I received from Baker at Los Banos was dated November 9, 1926. At that time he was still negotiating with Dr. Evermann, the director, and E. P. Van Duzee, the curator of Entomology, of the California Academy of Sciences at San Francisco, concerning the possible disposal of his collection to that institution and a sufficient salary to enable him to live and continue his work on his collection. His evaluation of the collection is expressed in the following words:

The collection is undoubtedly the largest existing private collection covering extreme Western Pacific. The prime part of it is contained in one thousand five hundred cases all crowded full. But as much more has been placed in hands of 110 specialists and considerable portions of the latter will be returned. Actually I believe it is one of the most important collections basic to either Central and South Pacific work or to S. W. Asian studies since it includes several thousand types and cotypes. Moreover more material is constantly coming in and I have so arranged it that continued collection on a large scale will be made after I leave here. I also have a lot of fine Australian material constantly coming in. Moreover I have taken the fullest advantage of exchanges with European museums and with individuals in this way securing a vast number of species I lacked, many of these being cotypes. As I told Van Duzee it has taken me fourteen years to build this up and it ought to go to a *safe* place at once. As I told him also if a staff of men were employed and given time to do the work and cover the same field they could not possibly approach the present accomplishment with an expenditure of $50, 000 in salaries and expenses.

There is still a larger aspect of the case recently up in the proposal of Muir and his associates to send me for five years to South Sea Islands. But the disposal and safe location of this great collection must be arranged in cooperation with the California Academy greatly to the advantage of the latter.

I hope all of these considerations will get to the right people and receive careful thought. America has not yet taken any large part in the Western Pacific work but now has opportunity to do so more effectively. I hope the opportunity will not be lost.

Failing to complete the negotiations with the California Academy of Sciences, Baker turned to Hawaii, where one of his Pomona College students, David L. Crawford, president of the University, led him finally to accept an offer from a strong combination of the Sugar Planters' Experiment Station, the Bishop Museum, the University of Hawaii, and other groups to conduct an extensive entomological South Sea survey in the Pacific Islands--over "Wallace's Trail." Accordingly, he presented his resignation to the University of the Philippines, to take effect in November, 1927. Arrangements were made for him to have headquarters at the University of Hawaii with President Crawford and to house his great collections of insects, fungi, and other plant materials in the Bishop Museum. On June 9, 1927, the Board of Regents of the University had passed a resolution appointing him professor of tropical agriculture and dean emeritus of the College of Agriculture of the University of the Philippines, and director emeritus of the Experiment Station, effective December 1, 1927.

All of the difficulties and worries that beset Baker during this last year undoubtedly broke his weakening spirit and finally even this courageous man with all his optimism, courage, and determination was forced to yield. His frail body could no longer withstand the inroads of disease and overwork, and his tired mind was shattered with disappointments and suffering known only to himself and one staunch friend who stood by him through those last bitter days. On the very consummation of the final changes he suddenly broke. He was rushed to the Sternberg General Hospital, and there, after an illness of two weeks, he died on July 21, 1927, aged fifty-five years. He was carefully tended to the last by one of his own former students, Captain Doctor Leon Gardner. Although his fatal ailment was said to be dysentery which had suddenly changed from a chronic to an acute stage, Welles (1927) remarks that "the doctors scarcely said whether it was malignant malaria or amoebic dysentery or tuberculosis to which he succumbed at last. "[14] His death came as a great shock, for it was generally supposed that he could endure and resist everything that came his way. It upset all of the well-laid plans for his immediate future and materially affected the large group who depended upon him for the natural history materials which were almost their very sustenance and existence.

It was the desire of the officials of the University of the Philippines to provide a suitable burial for Dean Baker on the campus of the institution to which he had given so many good years of his life. However, when his will was read it was found that this was not to be. Although most of his life had been devoted to his professional tasks and to work on his plant and insect collections, there had been a spark of romance in the little house on the hill. This and other intimate phases of Dean Baker's life were known to none save a very few of his most intimate friends. Of these Dr. Robert L. Pendleton, [15] a younger and more vigorous professor of tropical soils and agriculture at the university, became the most beloved and the most trusted. During these last difficult and heartbreaking days, he was the Good Samaritan upon whom Baker could rely to fulfill his every wish. He was the only one who ever got a clear look into the private life of Baker. Pendleton was also named Baker's executor and requested to care for those who had so lovingly and willingly made his life in the "little house on the Campus hill" more than a workshop and a place to live.

Dr. Pendleton has very kindly detailed to me in a letter of April 13, 1953, this more intimate side of Dean Baker's life. Dr. Pendleton writes:

Your letter of 7th April, asking about Dean Baker has just come in. I welcome this inquiry for it is high time that the records be put straight.

It was in 1917, when we were on our way out to India, that we first met Dean Baker. As you know, I'd been an amateur botanist, so since we had some weeks in Singapore, waiting for a ship to India, I went to the Botanical Gardens. There we found Baker busy. Having had six month's leave, he was spending it there. We visited him several times, and became friends. While we were still there he received a cable asking him to return to Los Banos and become Dean. Dr. Copeland had finally severed his connections.

From time to time, while we were in India, working for the Maharajah of Gwalior, we heard from Dean Baker. It was about January, 1923 that we heard that he had been able to find a spot for me on his staff . . . so he offered me the post of Professor of Soil Technology in the College of Agriculture.

I arrived in the Philippines in September, 1923, and on my way to the Campus stopped at Dean Baker's house at the edge of Los Banos, high above the ground on the posts of hard wood, and with the galvanized iron roof. Here he lived alone, and worked on his insect collections afternoons and *nights*. Distant some hundred feet or more, in the garden of bananas and fruit trees, was the kitchen and the living quarters of the Japanese family which were his servants. They cooked for him, and the husband did the garden work of the acre or more. The kitchen was connected to the house by a passage, really a bridge.

When I saw Dean Baker this first time after 1917, I noted a great change in him; he had aged greatly. But since he had been invalided home from the tropics many years before to teach at Pomona (College), he considered himself a person living on borrowed time; I did not think too much of it, and only later will be noted the connection of his condition with the great tragedy which had happened the year before.

It was the next year, 1924, that Dean Baker was trying to get entomologists from the United States Department of Agriculture to help with the problems in the college and the experiment station. *Quezon* would not consider the matter favorably, replying that to get these two men from Washington would just "be another link in the chain of bondage" to the United States.

It was a continual battle to get enough funds to keep the work going. Previous to my arrival in Los Banos the Experiment Station had been given a good tract of padiland. But when the President of the University was asked for a budget to run the Farm, he blew up, and replied that if he had thought that there would be needed a budget to run the land he would have taken good care that they would not have received the land in the first place! So to use the land, and keep it in shape for further experimental work, it was let out to local farmers to run on shares.

I could get no money for travel to see the soils of the Philippines, so Dean Baker got me jobs with sugar centrals and the Philippine Sugar Association. Then he would give me leave without pay to take those jobs, and so to see the soils of those districts at least.

It was in the summer of 1927 that he was quite ill. He would seldom admit that he was not well, and drove himself to get work done. He paid regularly a Cuban insect collector [Julian Valdez] whom he had brought to the Philippines. The fellow would get drunk and drew

much money, *but he could collect,* and fantastic stories were told
of his prowess in catching insects. Students, too, were hired to
catch insects. In the later years Dean Baker, himself, rarely went
collecting; it seemed to take most of his time to work up the collections brought to him. These he sorted and sent to specialists in various parts of the world. He also sent about half of his salary, every
month to various of these specialists, to enable them to live and to
work on his materials. At the time of his death he had a mailing
list of about 120 entomologists who were working on this material.

As you doubtless know, he had a hand-written check list on 3 x 5
cards, in trays on tables in his main work room. There were about
35,000 cards, and in some cases there were as m^ny as five refer-
ences on a single card.

This time, in late June, I think it was, Dean Baker asked me to
take him to the hospital in Manila. I drove him down in our car and
saw him properly cared for. He was there hardly a week, failing
fast. Chronic amoebic dysentery was probably the basic trouble. I
have forgotten what the physicians called it. In his delirium he kept
talking about the *enormous numbers of insects*.

It was a losing fight, and he passed away in about a week. I hastened back to Los Banos to see what papers and other particular
effects he had left. He had left me *no* personal instructions, nor had
ever mentioned anything about his affairs, except that he thought a
great deal of his brother, Ray Stannard. who was better known as
"David Grayson" and Dean Baker was to us much the lovable sort
of person that we knew from the books as David Grayson.

I found his affairs in perfect order. His will was there, and he
had specified that I was his first choice to handle the affairs of his
estate. We kept on the old Japanese man and wife to keep the collections dry and to look after the property.

The will clearly specified that his body was to be cremated, and
that the ashes were to be buried in a little Buddhist Temple yard in
Saimura, Tatsuno, about 50 miles northwest of Kobe (Japan). (In
the meantime the College of Agriculture authorities had made plans
for burial at Los Banos on the Campus but these plans could not be
followed.)

After the service in the Union Church, Manila, I went with the body
and saw to the cremation. The next morning I went with the under-
taker . . . to collect the ashes which were placed in a small narrow
box, and were kept in a safe deposit box in Manila until I could go
to Japan and carry out the rest of the clauses in the will.

It was in September, 1922, his cook at that time was Tome San,
a widow of about 36 years of age. The Japanese gardener seemed
to go insane and tried to kill Dean Baker with an axe. Tome San saw

the danger just in time, ran between Dean Baker and the gardener, and was so badly wounded with the axe in the hands of the gardener that she lived only five days, dying at the old Spanish hospital in town. Dean Baker was spared without even a scratch, but the blow was almost more than he could bear.

In his note he tells of his plans for marrying Tome San as soon as he could get his final divorce from his American wife. The divorce papers did not arrive until some years after the murder. Mrs. Baker and Charles Fuller Baker had not lived together for many years; as far as I recall, she did not go to the Orient with him at all.

Immediately after Tome San's death Dean Baker assumed her responsibilities. She had been supporting her niece in the Mission School, and had otherwise been looking after needs. The niece, Matsuko Matsumoto, had never been in the Philippines. But Dean Baker wrote her the most beautiful letters, as a father would to a daughter, and often wrote to the school principal, Miss Holland, also.

In addition to cremation the will called for the burial of the ashes beside those of Tome San and for the residual property of the estate to be given to Tome San's niece, an orphan student in the Palmore Institute (a Southern Methodist Mission Girls' School, Kobe, Japan).

In accordance with the will, I sent money regularly to the Palmore Institute for Mitsuko's schooling. Other bequests of the will were gradually completed.

His will also provided that his collections of insects were to go to the Smithsonian Institution and National Museum soon after his death. The Smithsonian Institution sent out Dr. [R. A.] Cushman, Specialist in the Ichneumonidae, to pack and ship the 1,400 (?) odd Schmidt boxes of insects[16] back to Washington, D.C. Also shipped were the 35,000 cards of the index of Malaysian insects, mentioned above. It was five months that Dr. Cushman lived with us at the college and worked daily on the repinning, packing, and shipping of the collections.

It was not until April, 1928, nine months later, that I was able to get leave from the college, and go to Japan with the ashes and see that they were buried. This was to be a personal job, for Dean Baker.

At the Palmore Institute I met the legatee, Mitsuko Matsumoto, who had just graduated from the high school course. With her help and that of the principal, Miss Holland, and the pastor of the Protestant church there, we arranged a funeral ceremony at the grave. It was a combined Buddhist and Christian service. As directed, the small casket of ashes was buried close to the stone marking Tome San's ashes.

Later we had a modest stone made to mark Dean Baker's grave.

Ray Stannard Baker suggested the inscription which in substance is: "He gave his life for the education of other people."

The temple garden, and the ancient structure itself, were most beautiful, and situated in the completely rural Japan of the pre-war days.

There were of course the formal dinner and the gifts to the temple for the upkeep of the grave. It was cherry blossom time! After a month's trip to Hokkaido I visited the temple once more.

In 1919 the Board of Regents of the University of the Philippines established the Baker Memorial Professorship in the College of Agriculture in memory of Charles Fuller Baker. This professorship provides for the services of men from abroad who will be in residence at the college for at least eight months and who will teach at least five hours a week.

ON THE HISTORY OF SCIENCE
AND OF THE SECOND LAW
OF THERMODYNAMICS

BY FREDERICK O. KOENIG

I. INTRODUCTION

§1. *Object of this article*

My object in the present article is twofold. In the first place I shall present a few simple ideas bearing on the history of science in general--ideas whose growth in my mind has been furthered not a little by the activity of our History of Science Dinner Club. I have found these ideas helpful in my attempts to teach the history of science, and I hope others may find them so. In the second place I shall present some of the history of the Second Law of thermodynamics. This presentation will be guided by the general ideas just mentioned and will illustrate them. As for the Second Law, it is conspicuous among the major generalizations of science for the variety of the forms in which it has been stated, as well as for the range of its implications and applications. Its history is correspondingly ramified, and therefore space alone would prevent me from giving here a complete account. I shall confine myself chiefly to the main features of what may be called the formative period of the Second Law, which begins in 1824 with the work of Sadi Carnot and ends in 1865 with the enunciation of the entropy concept by Clausius.

II. IDEAS ON THE HISTORY OF SCIENCE IN GENERAL

§2. *On the meaning of "science"*

Examination of the way in which the word "science" is currently used shows that it has not one meaning but several. Of these I wish to point out the following four: "Science" may mean (1) a certain kind of human activity; or it may mean (2) a certain body of knowledge, namely, all the knowledge called scientific; or it may mean (3) the union of (1) and (2); this is plainly the most comprehensive meaning of the word and that commonly intended when it occurs in the phrase, "the history of science"; or finally "science" may (and in the phrase "a science" always does) mean (4) any one of a set of well-recognized parts of (3), whereby any such part consists of a kind of activity (in turn a part of science under meaning 1) plus the associated body of knowledge (in turn a part of science under meaning 2); thus we speak of "the science of chemistry," or of "thermodynamics." These distinctions, although extremely obvious, are frequently overlooked, and discussions of science and of its history would benefit from closer attention to them. Meanings (1) and (2) I shall refer to as the principal meanings of "science."

Let me enlarge a little on my two principal meanings. First, with respect to meaning (1): the *activity* in question is always aimed at obtaining positive knowledge, regardless of whether or not this is done for an ulterior practical objective. Moreover, the activity is at least partly (usually largely) conscious, and in its conscious aspect is systematic rather than haphazard. Furthermore, it employs method (or if you like, methods) which involve observation, experiment, creative imagination, logic, and luck. Finally, it requires that the results found by an individual be communicated to and verified by others. With respect to meaning (2): the *body of knowledge* in question is mostly the result of the activity just described; it is cumulative; and it is logically ordered, at any time, as well as feasible--whence it is also said to be "organized," "integrated," or "systematized." These statements constitute in no wise a definition, but only a partial description.

§3. *On the unity of science*

It is above all George Sarton who has associated with the history of science the notion of its unity. What now, in "the unity of science"--"this heartening phrase" as Robert Oppenheimer has recently called it--is the meaning of the word "science"? I submit that it has either of its two principal mean-

ings as described above, and that "the unity of science" also
receives two alternative meanings accordingly.

Under science, the activity (meaning 1), the unity of science
has been described succinctly and eloquently by Oppenheimer,
and we cannot do better than quote him. He calls it "a unity of
comparable dedication, " and elucidates this as follows: "The
open society, the unrestricted access to knowledge, the un-
planned and uninhibited association of men for its furtherance--
these are what may make a vast, complex, ever-growing,
ever-changing, ever more specialized and expert technological
world, nevertheless a world of human community. "[1]

Under science, the body of knowledge (meaning 2), the unity
of science is merely the fact already mentioned, that this
knowledge is logically ordered as well as feasible. This does
not mean that if the ordering should remain, as seems most
likely, forever imperfect, then science would be a failure. Nor
does it suggest that any individual could comprehend the entire
scientific corpus as a unit, or should try to. What it does mean,
for instance, is that any scientist is likely to profit from ac-
quaintance with fields adjacent to his specialty; and that we
shall continue to have delight and other profit from the great
explanations achieved in the past, such as those of Newton,
Darwin, Maxwell, or Einstein; and that the striving to create
a unitary theory comprising electromagnetics and the various
branches of mechanics--relativistic, classical, quantum, and
nuclear--will continue, as will that to interpret biology and
psychology in physicochemical terms.

§4. *Inner and outer history of science*

I have indicated that in the term "the history of science, " we
take the word "science" in its most comprehensive sense,
which is the sum of the two principal meanings. I now suggest
that for the study of the history of science, thus understood,
it is helpful to divide this history into two parts, which I shall
call, for convenience, "the inner" and "the outer history of
science. " For clarity let me say immediately that although
there is, as we shall see, a kind of correspondence between
on the one hand these two divisions of the history of science,
and on the other the two principal meanings of "science, " yet
the distinction between the inner and outer history of science
is not made by inserting in turn the two principal meanings

into the phrase "the history of science. " Actually the division
is defined as follows: The *inner history of science* deals with
the chronology of the development of science (meaning 3) and
with the relation of this chronology to the logical order (at all
times imperfect) of the body of scientific knowledge; and it
does this, moreover, with a minimum of reference to extrascientific matters. Then necessarily, the *outer history of
science* deals historically with the interrelations between
science (meaning 3) and the main bulk of extrascientific matters.

Let us consider for a moment the inner history of science.
Evidently it can be studied independently of the outer. As obvious examples of the sort of thing to be learned from such
study I shall cite the following: there could be no celestial
mechanics before Copernicus' discovery of the appropriate
frame of reference; hardly a satisfactory chemistry before a
correct pneumatics; neither a Mendeléeff nor a Kekulé before
an Avogadro; hardly an absolute temperature scale before a
scientific analysis of heat engines; no electromagnetic theory
of light under a purely particulate theory of optics; no theory
of evolution before a correct paleontology. Plainly, the inner
history of science pursued independently of the outer gives
valuable emphasis to the unity of science in both of its aspects
(§3), but more particularly in that aspect corresponding to
science, the body of knowledge (meaning 2).

Now consider the outer history of science. Its function is
to display the history of science as the integral part which the
latter, of course, constitutes of history as a whole. However,
the outer history cannot be discussed without at least some
reference to the inner, so that the relation between the two
is unsymmetrical. Among the things belonging specifically to
the outer history I shall mention especially the following: the
biographies of the scientists; the historical treatment of the
interactions between science, on the one hand, and technology,
economics, and politics on the other--interactions which have
come to be known as "the social relations of science"; such
parts of the histories of other fields of culture, above all religion, philosophy, literature, the fine arts, and music, as are
related to the history of science. It thus appears that the outer
history comprises all the problems about those manifold

causes, or motives, of scientific activity that lie outside of science itself. These problems include some of the most important and difficult of the entire history of science, as for instance: What factors account for the unique development of the experimental method in Western Europe? What is the relation between scientific achievement and personality, and how does this relation change with changing cultural climate? Plainly, the outer history of science gives valuable emphasis to the unity of science in both of its aspects (§3), but more particularly in that aspect corresponding to science, the activity (meaning 1). If now we compare this last sentence with the last sentence of the preceding paragraph, we see the correspondence I have mentioned as subsisting between, on the one hand, the two divisions of the history of science and, on the other, the two principal meanings of "science."

Finally, I must point out that the division into an inner and an outer part applies to the history not only of science as a whole, but also of any of its recognized fields--whether or not such a field is ordinarily said to be "a science" (meaning 4). This will be illustrated by the remainder of the present article, in which I shall consider separately the inner and outer histories of the Second Law of thermodynamics. The logical order would be to take the inner history first. I shall, however, reverse this order for the benefit of those readers--undoubtedly a majority--who are more interested in history than in thermodynamics.

III. ON THE HISTORY OF THE SECOND LAW

A. Outer History
§5. *Genesis of the Second Law*
The genesis of the Second Law of thermodynamics is definite to a degree that is almost embryological and that is therefore rare in the history of science. The Second Law was conceived at the moment that Sadi Carnot (1796-1832) began consciously to act upon the impulse to apply methods of science to the steam engine. As to the precise date of this event, and thus as to the time of gestation of the new being, I can say nothing. [2] The birth occurred in 1824, with the publication by Sadi Carnot of a pamphlet of 118 pages entitled *Réflexions sur la puissance*

motrice du feu et sur les machines propres a développer cette puissance.[3]

The *Réflexions,* though in form an essay, i.e., a single chapter without subheadings, falls into a number of readily distinguishable parts. The first of these (pp. 1-9), of an introductory nature, has a strong bearing on the genesis of the Second Law, and I shall therefore discuss it immediately. The main bulk of the work (pp. 9-118) I shall discuss later, in connection with the inner history.

Carnot begins by pointing out that heat may be a cause of motion, as evidenced by the steam engine and also by the "vast disturbances" occurring on the earth, e.g., wind, rain, and volcanic eruptions. He goes on to review the manifold uses of the steam engine, among which he emphasizes particularly coal mining and "the safe and rapid navigation by means of steamships." There follow two paragraphs on the history of the steam engine, of which the first is worth quoting here.

The discovery of the steam-engine, like most human inventions, owes its birth to crude attempts which have been attributed to various persons and of which the real author is not known. The principal discovery consists indeed less in these first trials than in the successive improvements which have brought it to its present perfection. There is almost as great a difference between the first structures where expansive force was developed and the actual steam-engine as there is between the first raft ever constructed and a man-of-war.

Hereupon Carnot states the main object of his work as follows:

In spite of labor of all sorts expended on the steam-engine, and in spite of the perfection to which it has been brought, its theory is very little advanced, and the attempts to better this state of affairs have thus far been directed almost at random.

The question has often been raised whether the motive power of heat is limited or not; whether there is a limit to the possible improvements of the steam-engine which, in the nature of the case, cannot be passed by any means; or if, on the other hand, these improvements are capable of indefinite extension. Inventors have tried for a long time, and are still trying, to find whether there is not a more efficient agent than water by which to develop the motive power of heat; whether, for example, atmospheric air does not offer great advantages in this respect. We propose to submit these questions to a critical examination.

Carnot then concludes his introduction with a brief and profound forecast of a great part of the program of the science, as yet unnamed, of thermodynamics.

The phenomenon of the production of motion by heat has not been considered in a sufficiently general way. . . .
The machines which are not worked by heat--for instance, those worked by men or animals, by waterfalls, or by air-currents--can be studied to their last details by the principles of mechanics. . . . The theory of such machines is complete. Such a theory is evidently lacking for heat-engines. We shall never possess it until the laws of physics are so extended and generalized as to make known in advance all the effects of heat acting in a definite way on any body whatsoever.

Thus Carnot begins by asking certain questions about a practical device which had been developed largely without the help of science, and he does so with an eye to the possible improvement of this device. These questions, e. g., "whether the motive power of heat is limited or not, " etc., are very concrete and at the same time wonderfully general. The answers that Carnot gets constitute, as is well known, the first and greatest step in the discovery of the Second Law. Moreover, even after the completion of the discovery by Clausius (1850) and by Thomson (1851-52), it was a considerable time before its immediate consequence--the science of thermodynamics--began to influence (as anticipated by Carnot) the design of steam engines. Hence the work of Carnot is an example of the fructification of science by technology--remarkable, I may add, for its association with a single individual, its directness, and its grandeur of scale. All this is, of course, nothing new. Professor Lawrence J. Henderson (1878-1942), who initiated the study of the history of science at Harvard, was fond of saying that "the steam engine did much more for science than science ever did for the steam engine. "

I must, however, warn against the oversimplification of regarding the steam engine of 1824 as the creation of a series of "sooty empirics" (an epithet of Boyle's) completely innocent of science. Actually the development occurred not entirely without benefit from science, but only largely so, as I have said. Of the pioneers of the steam engine at least two were very close to science: Denis Papin (1647-1712) and James Watt (1736-1819).

Papin was an assistant of Huygens (1629-95) and later of Boyle (1627-91). And Huygens at one time designed (though did not build) an engine to be operated by gunpowder, whereas Boyle was concerned with the air pump, a "philosophical" device suggestive of the steam-engine-to-be. As for Watt, in his early days as an instrument maker in Glasgow, he learned from Black (1728-99) of the latter's discovery of latent heat, and later he applied this knowledge in his most important invention, the condenser, patented in 1769.

Having arrived at this point, I wish I could display the birth of the Second Law from the steam engine in the perspective of the whole history of the relation between science and technology, I shall, however, only say that owing to the vast growth, since the time of Carnot, of science, of technology, and of the degree of coupling between them, a contribution to science in the future from some field of practice nearly independent of science--particularly a contribution on the scale of the Second Law--seems unlikely. As for the relation between science and technology in general, its history is a subject so novel and complex that it is no wonder we have as yet no adequate treatment.[4] It has of course been used by the Marxists, who thus did the service, I feel constrained to admit, of calling attention, before Hiroshima, to the urgency of the problem.

§6. *Biographies of the three discoverers*

I have considered the "birth" of the Second Law as embodied in the *Réflexions* of Carnot, and have indicated that I shall distinguish this from its "discovery, " which is due to three men: Carnot, Clausius, and Thomson. What I mean here by "discovery" will appear below (§8), in my discussion of the inner history of the Second Law. Before coming to this, I shall end my account of the outer history, well aware of its incompleteness, with a biographical sketch of each of the three discoverers.

a. Carnot (1796-1832)

Nicolas Léonard Sadi Carnot was born in Paris in the Year IV. The Carnot family is a striking example of hereditary genius, reminiscent of the Darwins. The outstanding members are, in addition to Sadi, his father Lazare Nicolas Marguerite (1753-1823), of whom I shall say more in a moment; his younger brother Lazare Hippolyte (1801-88), distinguished as a writer and statesman; and finally the two sons of the latter, Marie Francois Sadi (1837-94) and Marie Adolphe (1839-after 1882), who became respectively a president of the Third Republic and a chemist of note. Sadi's brief career was so critically influenced by that of his father that I must introduce a sketch of the latter.[5]

Lazare Carnot is unique in that he took a more active part by far in the great and terrible events of the French Revolution and the Napoleonic period than any other person of equal or greater stature as a scientist. He was born in the village of Nolay to a middle-class family of long standing and in 1773 began his career as a military engineer. Projected into the storm center of the Revolution in 1791 by election to the Legislative Assembly, he quickly rose to high prominence as a military and political leader and in 1795 became one of the five Directors. It was during his occupancy of this post that his elder son, Sadi, was born (June 16, 1796).

In 1797 Lazare had to flee the country. In 1799, under the Consulate, he returned and in 1800 became minister of war. However, being strongly opposed to Napoleon's course of self-aggrandizement, Lazare in 1801 resigned this post, went into retirement, and devoted himself largely to study. Nevertheless, in 1814 patriotism overcame his republican scruples, and he returned to military service as governor of Antwerp. During the Hundred Days, Napoleon made Lazare minister of the interior and a peer of France. Consequently after the Second Restoration, he was proscribed and went into exile. He spent the last seven years of his life in Magdeburg.

Lazare Carnot's record is adorned, from 1783 to 1820, by a series of noteworthy contributions to mathematics, mechanics, and the art of fortification, and by a volume of minor poetry. Of these various works one has a direct bearing on our present purpose. This is the principal work on mechanics, which appeared in 1803, entitled *Principes fondamentaux de l'équilibre et du mouvement.*[6] Here Lazare Carnot discusses, among other things, the efficiency of purely mechanical devices in a way that is suggestive of the approach of his son Sadi to the analysis of the steam engine. I shall revert to this point in §9h. His work on mechanics, however, is by no means Lazare's greatest contribution to science. This place is filled by the *Géométrie de position* of 1803, [7] a highly original work on pure mathematics, which became one of the foundation stones of the present science of projective geometry.

Finally, Lazare Carnot had a hand in two of the achievements of the French Revolution relating to the organization of scientific activity: (1) the foundation in 1794 of the École Polytechnique, destined to be the cradle of so many great scientists, including Sadi Carnot, and (2) the foundation in 1795 of the Institut National, within which the Académie des Sciences, suppressed in 1793, was revived.

Returning now to Sadi Carnot, [8] we find a career that in contrast to his father's long and spectacular one is short, unobtrusive, and vastly more important for science. We are told that once at the age of five he ran away and was found "a long way off, in a mill, the mechanism of which he was trying to discover. " This early curiosity concerning machines--a trait, incidentally, of many of the great scientists, [9] including Newton--led Carnot *père* "to give a scientific direction to the studies of his son. " As a prospective military engineer Sadi attended the aforementioned École Polytechnique, from 1812 to 1814. In March of the latter year he had a brief taste, his only one, of actual fighting, when he took part in the futile defense of Paris against the Allies. In October, 1814, he left the Polytechnique for a school of military engineers at Metz, where during the Hundred Days he obtained

. . . a glimpse of human nature of which he could not speak without disgust. His little sub-lieutenant's room was visited by certain superior officers who did not disdain to mount to the third floor to pay their respects to the son of the new minister. Waterloo put an end to their attentions . . . and Sadi [was] sent successively to many trying places to pursue his vocation of engineer, to count bricks, to repair walls, and to draw plans destined to be hidden in portfolios. He performed these duties . . . without hope of recompense, for his name, which not long before had brought him so many flatteries, was henceforth the cause of his advancement being long delayed.

By 1818 Carnot's frustration had reached a point where he welcomed an opportunity, offered by a royal decree affecting all army officers, to go on permanent leave while remaining attached to the service with the rank of lieutenant. Thereupon he threw himself into a life of ardent study which he continued,

Sadi Carnot, six years after the publication of the *Réflexions*

in and near Paris, until his death in 1832, with but two significant interruptions. The first of these was a journey to Magdeburg in the summer of 1821, to visit his exiled father. The second, more extended, began in 1826 when Carnot was recalled to active duty with the rank of captain, and ended in 1828 when, finding military life oppressive in spite of promotion, he resigned from the service completely.

As a private scholar from 1818 on Carnot subjected himself to a curriculum comprising mathematics, physics, chemistry, natural history, parts of technology, economics, French literature, and music. In these pursuits he made use of the opportunities offered by the Sorbonne and other institutions of higher learning in Paris, as well as the museums, factories, and theaters. For relaxation he cultivated a variety of sports such as gymnastics, fencing, and dancing. This diverse activity, albeit intense, served only as the matrix for Carnot's creative work--the investigation of heat and its motive power.

Although "in small companies . . . not at all taciturn, " Carnot "had such a repugnance to bringing himself forward that, in his intimate conversations with a few friends, he kept them ignorant of the treasures of science which he had accumulated. They never knew more than a small part of them. " Our knowledge of these treasures is therefore derived solely from Carnot's published writings, and these unfortunately shed almost no light on the evolution of the ideas in his mind. These writings comprise in all two items, of which only one was issued by Sadi Carnot himself, the *Réflexions,* in 1824. The other item consists of a series of notes for his private use, which became known to the world only when in 1878, forty-six years after Sadi's death, his brother Hippolyte published selected portions as an appendix to the second edition of the *Réflexions.* [10] These notes prove that among the "treasures" kept back by Carnot was the equivalence of heat and work. Thus had he not died prematurely, thermodynamics would probably have advanced more rapidly than it did. Exhausted by overwork Carnot fell ill of "an inflammation of the lungs, followed by scarlet fever, " and finally, in August, 1832, succumbed to cholera, which was then sweeping Paris.

   b. Clausius (1822-88)

   Rudolf Julius Emmanuel Clausius was born in Köslin, Pomerania. [11] He became a mathematical physicist, and his *curriculum vitae* is typical of that of the German professors of his time. After attending the Gymnasium at Stettin, he studied at the University of Berlin from 1840 to 1844. In 1847 he published in *Crelles Journal* his first scientific paper, dealing with the scattering and reflection of sunlight by the atmosphere. He became an instructor in physics at the Royal Artillery and Engineering School in Berlin and *Privatdozent* at the University in 1850. The same year witnessed the publication of his first paper on the theory of heat, [12] which is also his greatest because it contains his chief contribution to the discovery of the Second Law (see §11).

   How and when Clausius' interest in the relation between heat and work got started, I have been unable to find out. In the paper in question the ideas of Carnot play a leading role (along with those of Joule), but Clausius obtained his knowledge of those ideas only indirectly. Referring to the *Réflexions* he says, "I have not been able to obtain a copy of this book, and am acquainted with it only through the work of Clapeyron

Clausius in his Zurich period (1855-67)

and Thomson. . . . " Clapeyron's work, [13] the first after the *Réflexions* itself to apply the ideas of Carnot, had appeared in 1834 (see §10a). The work of Thomson, constituting Clausius' other source, was a paper of 1849[14] (see the biographical sketch of Thomson, below). After his paper of 1850 Clausius persevered in investigating the fundamental aspects of the new law. The result was a series of papers, which extended to 1867 and reached a triumphant climax in 1865 with the definition of the entropy (see §15b).

In 1855 Clausius was called to Zürich as professor *(ordinarius)* of physics at the Polytechnicum and in 1857 was also appointed professor at the university. He held both posts until 1867 when he moved to Würzburg as professor of physics. In 1869 he took the chair of physics at Bonn, which he occupied for the remainder of his life. This sober progression was interrupted briefly by the Franco-Prussian War of 1870-71, in which he served as head of an ambulance corps of Bonn students.

Second in importance to Clausius' work on thermodynamics

is his contribution to the closely related subject of the kinetic theory of gases that begins in 1857 with a paper entitled, "Über die Art der Bewegung, welche wir Wärme nennen." These investigations paved the way for the creation of statistical mechanics by Maxwell and Boltzmann. In addition, Clausius enriched the fields of electrodynamics, potential theory, and electrolysis.

Gibbs[15] estimates the total number of Clausius' scientific papers as about 115. All the papers on thermodynamics together with all those on kinetic theory, and with a few on electricity in relation to thermodynamics, were republished by Clausius in *Abhandlungen über die mechanische Wärmetheorie*. His work is characterized throughout by a striving for generality and systematization, which, along with the uniformity of his outward career, suggests the "classic type of genius" defined by Ostwald.[16]

c. Thomson (1824-1907)

William Thomson, better known under the title of his later years, Lord Kelvin, was one of the greatest physicists of his century and one of its commanding scientific personalities. He was born in Belfast to a North Irish family of Scottish extraction.[17] The father, James Thomson, had risen from small beginnings to be a teacher of mathematics at the Royal Academical Institution in Belfast. William was the fourth child (and second son) of a total of seven (of whom four were boys). His older brother James (1822-92) became a distinguished physicist and engineer and contributed to the development of the Second Law; I shall return to his work below (§10e). In 1830 the mother died. Her role with the children was taken over by the father who did not marry again. In 1832 the family moved to Glasgow, where the father had been appointed professor of mathematics at the University.

The boys James and William never attended school and received all their elementary instruction from their father. Both boys, but especially William, were prodigies. They enrolled in the University of Glasgow in 1834, William being then ten years old and James twelve. In the summer of 1839 the family visited France to learn the language, which William did by reading Laplace's *Mécanique céleste*. Within the next year he mastered Fourier's *Théorie analytique de la chaleur* so well that, during a visit to Germany in the summer of 1840 for the purpose of learning German, he wrote a paper correcting erroneous views on Fourier that had been published by Kelland, the professor of mathematics at Edinburgh. This paper, Thomson's first, was entitled "On Fourier's Expansions of Functions in Trigonometrical Series" and appeared in the *Cambridge Mathematical Journal,* under the pseudonym "P. Q. R.," in 1841.

After this, every year through 1908 witnessed the publication of papers by Thomson. The total listed by S. P. Thompson[18] is 661, which makes an average, over the 68 years in question, of 9.72 papers per year. Of these papers a few are of the first importance scientifically, many are very important, and none is trivial or repetitious. The record is made even more astounding by Thomson's complete or partial authorship (amounting in some cases only to the preface) of twenty-four books (1863 to 1906), and by his seventy patents (1854 to 1907).

The year 1841 also saw Thomson's entrance into Peterhouse College, Cambridge. He graduated in 1845 as Second Wrangler and First Smith's Prizeman. Early in 1846 he spent four and a half months in Paris, for training in experimental technique in the laboratory of H. V. Regnault (1810-78), who was then engaged in his classical researches on the thermal properties of gases and vapors, undertaken with the support of the French government for the express purpose of improving the steam engine. Stimulated probably by this contact, Thomson read Clapeyron's paper of 1834 ("Sur la puissance motrice de la chaleur") mentioned above. He thereupon, while in Paris, tried to obtain a copy of the *Réflexions* but was unsuccessful. Thus Thomson, like Clausius later on, obtained his first knowledge of Carnot indirectly. Thomson first saw the *Réflexions* late in 1848, after his discovery of the absolute temperature (see §10d).

In September, 1846, he was elected, at the age of twenty-two, to the professorship of natural philosophy at Glasgow. He thereby fulfilled one of the dearest ambitions of his father. He remained in the Glasgow post for the next fifty-three years, resigning from active duty in 1899.

In 1847 at a meeting of the British Association he met Joule (1818-89), who had published his experimental discovery of the equivalence of heat and work in 1843, and who was then en-

William Thomson in 1852, at the height of his thermodynamic period.

gaged in improving his methods. This meeting dates the beginning of Thomson's overt research into the relation between heat and work; however, the lead he took up was not Joule's, but Carnot's as transmitted by Clapeyron. The first result was the above-mentioned discovery of the absolute temperature, in 1848.[19] This was followed in 1849 by his first paper based on direct knowledge of the *Réflexions* and entitled, "An Account of Carnot's Theory of the Motive Power of Heat, with Numerical Results Deduced from Regnault's Experiments on Steam."[20] Although containing nothing essentially new, this paper is important as the source for Clausius mentioned above and is also of interest in connection with the origin of the word "thermodynamics" (see §14).

Thermodynamics remained the center of Thomson's scientific attention for the next five years, and after this efflorescence he continued practically to the end of his long life to publish on various aspects of the subject. To list all these thermodynamic papers is not necessary here. The one most important for the discovery of the Second Law (see §12) appeared in 1851, [21] and the next most important (see §13) in 1852. [22] The classical researches with Joule, on thermal effects in fluids moving through small apertures, were published in the decade 1852-62, and the first paper on the thermodynamics of thermoelectricity appeared in 1854. [23] Thomson's thermodynamic work is his greatest contribution to science, and here I can mention only briefly some of the others.

In electromagnetics he adumbrated the discoveries of Maxwell and Hertz, greatly refined existing instruments of measurement and invented new ones, and furthered the systematization of the units. He enlivened geology by his erroneous hypothesis, based on consideration of the heat of the sun, of the age of the earth. He stimulated atomic theory by his erroneous hypothesis of the vortex atom. By his unwavering and powerful effort to apply universally the macroscopic point of view, he helped his successors to see its insufficiency.

From 1855 on--he was then thirty-one years old and had his most important contributions to science already behind him--Thomson became increasingly drawn into technology. This development began with his work for the Atlantic cable. The success of this great enterprise, achieved in 1866 after a decade of trial, was directly due to Thomson's theory of signaling and to the instruments he was thereby led to invent. Later he greatly improved the magnetic compass and the methods of sounding. In consequence he became famous and wealthy.

He was knighted in 1866 and in 1892 raised to the peerage as Baron Kelvin of Largs. He traveled extensively, after 1870 mostly in his own sailing yacht. He was married in 1852 and, after his wife died in 1870, for a second time in 1874. He had a friendship with Helmholtz (1822-94) from 1855 to the latter's death. The jubilee of his Glasgow professorship, in 1896, occasioned the gathering of two thousand scientists and others from many parts of the world. He is buried in Westminster Abbey. Thomson's precocity, the many-sided and sometimes fragmentary nature of his work, his change from pure to applied science in middle life, and his rise to fame, make him clearly an example of the "romantic type of genius" defined by Ostwald. [24]

B. Inner History

§7. *Features of the Second Law relevant to its inner history*
In order to appreciate the inner history of any field of science, one needs some knowledge of its present condition. The Second Law of thermodynamics is a good illustration. I have already mentioned (§1) that it is extraordinary for the variety of its statements. To understand the different ways in which the three men, Carnot, Clausius, and Thomson, contributed to the discovery of the law, we need to know some of these statements and also certain facts concerning them. I shall accordingly summarize, with a minimum of reference to history, this necessary scientific information.

a. Reversibility and some related concepts
A majority of the statements of the Second Law involve reference to reversible or irreversible processes. Although this concept is among the most familiar of thermodynamics, the definitions given in the literature are conflicting, in that some authors take the defining characteristic of a reversible process to be equilibrium, some take it to be restorability, and some use the term "quasistatic" as a synonym for "reversible." For clarity it will therefore be necessary for me to summarize my view of the matter.

I begin by confining the discussion to macroscopic systems for the scientific study of whose macroscopic behavior the laws of mechanics and electromagnetics are not sufficient. I define a *passage* as the traversing, by a system, over a period of time, of a sequence of its macroscopic states, whereby I include the fictitious case in which the traversing occurs infinitely slowly. I define a *process* as any passage, or any other macroscopic happening (e. g., flow of heat) which occurs over a period of time, and which is causally linked with a passage. I define any infinitely slow process to be *quasistatic*, if the states traversed in the passage that constitutes, or is causally linked with, the process in question are states of rest (i. e., states continuing unchanged in isolated systems). I define a process to be *possible* or *impossible*, according as it does or does not represent actual experience, whereby I take "represent" to permit the fiction of quasistatic process. I call any possible process, which is not quasistatic, a *natural* process. I assume for any possible process that the system in which it occurs can be brought into contact--while just that process is occurring--with other systems such that the combined system constitutes an isolated system, and I call the passage then made by this isolated system, a *global* process that *includes* the process in question. Finally, I define any possible process to be *reversible* or *irreversible,* according as there is or is not a global process including it such that, after that global process has occurred, the isolated system which has undergone that global process is capable of another global process which restores it (the isolated system and therefore the system in which the process in question has occurred) to its original state. Thus the defining feature of reversibility is restorability. For clarity I point out that not all quasistatic processes are reversible (e. g., quasistatic deformation of a solid beyond its elastic limit is irreversible) and further, that not all quasistatic processes are possible (e. g., the exact opposite of the quasistatic deformation just mentioned is impossible).

  b. The Generalized Second Law of thermodynamics
  In terms of the classes of process just defined, a vast body of empirical experience can be exactly summarized in the following form: *All natural processes, and some quasistatic ones, are irreversible.* This is, as far as I know, the most general statement possible within the context of thermodynamics and implying everything that has long been regarded as the characteristic content of the Second Law. Accordingly I hold that it is this statement or an equivalent one that *ought* to be taken as "*the* Second Law" in any modern exposition. This course, however, has not yet been generally adopted; most current and past expositions take as "the Second Law" one or another member of a class of statements less general than the above and deducible from it. Hence--my primary concern here being not the improvement of thermodynamics but the understanding of its history--I shall adhere to the traditional and still customary méaning of the term "the Second Law" and speak of the above more general statement as the *Generalized Second Law* (abbreviated: the G-Law).

  c. The principal statements of the Second Law
  I now turn to the class of statements just mentioned, whose members are customarily taken as the various alternative forms of the Second Law. The statements here given are my

own version of what the literature contains; if any of them should be exact quotations this is mildly accidental. I group the statements under seven headings each of which comprises a single statement except (6), which comprises a class. To each heading I give a name. My names for (1) to (5) inclusive and (7) are respectively those assigned in the literature to statements of identical or nearly identical content and are therefore familiar. When a statement is named after a person, e. g., Carnot's theorem, this means that its content was discovered wholly or largely by that person but (as already indicated) not that he expressed his discovery in the words used here.

(1) *The Clausius statement*

It is impossible to construct a device whose sole effect is the transfer of heat from the cooler of two systems at different temperatures to the hotter.

(2) *The Kelvin statement*

It is impossible to construct a device whose sole effect is the extraction of a quantity of heat from a system of uniform temperature, and the performance of an equal quantity of work, or more briefly: *A perpetual motion machine of the second kind is impossible.*

(3) *Carnot's theorem*

Call a heat engine *simple* if its thermal interaction with its surroundings consists only in the absorption of heat from a reservoir at a fixed temperature and the rejection of heat to another reservoir at another fixed temperature; and call any simple engine which is reversible a *Carnot engine.* Then Carnot's theorem is the following: (a) *All Carnot engines between the same two temperatures have the same efficiency;* (b) *if the efficiencies of two simple engines between given temperatures are equal, and one of these is a Carnot engine, then so is the other;* (c) *the efficiency of an irreversible simple engine between two temperatures is less than that of a Carnot engine between those temperatures.*

(4) *Clausius' theorem*

Call a system *closed* if matter neither enters nor leaves it (from or to the surroundings). Then Clausius' theorem is the following: *For a cycle undergone by a closed system in which the latter receives the (positive or negative) quantities of heat*

$q_1, q_2 \ldots q_n$ *from heat reservoirs respectively at the ab-solute temperatures* $T_1, T_2 \ldots T_n$, *it is true that*

$$\sum_{i=1}^{n} \frac{q_i}{T_i} \begin{cases} =0, \\ <0, \end{cases} \quad \begin{matrix} \textit{if the cycle is reversible;} \\ \textit{if the cycle is irreversible.} \end{matrix}$$

(5) *The entropy principle*

Consider a closed system having states, Z, which are connectible by reversible passages and also by irreversible ones. Consider also an infinite set of heat reservoirs whose respective absolute temperatures are distributed continuously over the entire temperature range and which serve as the sole sources of such heat as the system receives (positively or negatively) in any passage. We denote by T the absolute temperature of any reservoir furnishing the element of heat dq to the system. Then the entropy principle is the following: (a) *If we select any state,* Z_0, *as a reference state, then the system has in each state Z a macroscopic property--the entropy, S--given by*

$$S = S_o + \int_{Z_o \overset{\text{rev}}{\longrightarrow} Z} \frac{dq}{T},$$

where $Z_0 \overset{rev}{\longrightarrow} Z$ *denotes any reversible passage from* Z_0 *to* Z, *and* S_0 *is an arbitrary constant;* (b) *for any irreversible passage connecting two states,* Z_1 *and* Z_2:

$$S(Z_2) - S(Z_1) > \int_{Z_1 \overset{\text{irrev}}{\longrightarrow} Z_2} \frac{dq}{T}$$

(6) *The irreversibility statements*

There exists an indefinitely large class of processes, the declaration of any member of which to be irreversible constitutes a statement of the Second Law. Examples of such processes are: (a) the flow of heat down a finite temperature gradient; (b) the dissipation of work by friction; (c) the free expansion of a gas; (d) the explosion of a mixture of hydrogen and oxygen in a rigid adiabatic container. Denote any member of this class of processes by N. Then the irreversibility statements of the Second Law are subsumed as follows: *N is an irreversible process.* For instance, taking N to be process (c) of the set of examples just given, we get: *the free expansion of a gas is an irreversible process;* and this is one statement of the Second Law. I may add that the statements obtained from processes (a) and (b) are already seen to be equivalent respectively to the Clausius statement and the Kelvin statement given above. These statements of the Second Law are not as well known as they ought to be, and the name "irreversibility statements" represents only my own suggestion.

(7) *The Carathéodory statement*

For a closed system, any accessible state of rest has neighboring states of rest infinitesimally removed from the given state and inaccessible from it by any kind of adiabatic path.

    d. Logical relation of the various statements of the Second Law to the G-Law and to one another; role of the First Law

For the various statements of the Second Law in §7c it can be shown that if we take as premises any one of these *and in addition the First Law of thermodynamics,* then we can deduce all the other statements of §7c; or in short: *under the First Law, the various statements of the Second Law given in* §7c *are mutually equivalent.* This fact played a crucial part in the discovery of the Second Law, as we shall see. We may add that statements (1), (2), and (6) of §7c follow very readily from the G-Law (§7b). [25]

§8. *The discovery of the Second Law, in summary*

We are now ready to consider the inner history of the Second Law. I shall begin by emphasizing that the statements of §7c, which set forth the Second Law as currently understood, are in certain minor respects more complete than most of the statements given by the discoverers, Carnot, Clausius, and Thomson. The differences represent a labor of refinement carried out over the hundred years and more since the discovery. This is particularly true in the case of Carnot. To say, "Carnot discovered Carnot's theorem, " if by "Carnot's theorem" we mean the modern version given in §7c, is an oversimplification too coarse for good history. In order to deal with this situation without being drawn into a discussion of details too lengthy for the present purpose, I shall take the phrase, *the essential content of x,* to denote any physical meaning which is either identical or nearly identical with the physical meaning of *x,* where *x* in turn denotes any of the statements of §7c. I shall furthermore abbreviate the phrase just defined in the form, $x(e.c.)$. Thus, it will now be correct to say, "Carnot discovered Carnot's theorem (e. c.), " meaning, "Carnot discovered the essential content of Carnot's theorem as stated in §7c."

I shall now summarize the events that I regard as constituting the discovery of the Second Law. First, Carnot in 1824 *(Réflexions)* discovered Carnot's theorem (e. c.) by deducing it from certain premises (see §9); originated the concept of reversibility; demonstrated the fertility of Carnot's theorem (e. c.) as a premise for further deduction in the theory of heat. Second, Clausius in 1850 ("Über die bewegende Kraft der Wärme") enunciated the Clausius statement (e. c.); deduced Carnot's theorem (e. c.) from the Clausius statement (e. c.) and the First Law as joint premises; recognized the existence of a Second Law as distinct from a First Law; obtained the first empirical verification of both laws jointly. Third, Thomson in 1851 ("On the Dynamical Theory of Heat") enunciated the Kelvin statement (e. c.); deduced Carnot's theorem (e. c.) from the Kelvin statement and the First Law as joint premises; pointed out the equivalence, under the First Law, of the Kelvin statement (e. c.) and the Clausius statement (e. c.); and in 1852 ("On a Universal Tendency in Nature") adumbrated the G-Law. Thus the discovery of the Second Law occurred in two bursts, in 1824 and 1850-51-52 respectively; and within the interval between these the First Law was discovered. This summary will orient us throughout the more detailed account which follows.

§9. *Carnot's Réflexions*

a. General scheme

I have discussed the introductory part of the *Réflexions* (pp. 1-9) in §5, and turn now to the main bulk of that work. As its title suggests, the work contains no experiments by Carnot. Its method consists in assuming a set of premises derived from inspection of the steam engine and from current ideas, in deducing from these premises certain consequences, and finally in comparing some of the latter with available empirical data. The chief premises are (1) that the motive power of heat is associated with differences of temperature, and conversely; (2) that heat is a kind of weightless fluid which is conserved in all processes; (3) that perpetual motion is impossible. Premise (2), which is of course false, I shall call

the *caloric theory*. From these premises is deduced immediately the most important result of the work, namely, Carnot's theorem (e. c.). Hereupon, the *Réflexions* proceeds to
use Carnot's theorem (e. c.) as a premise for further deduction, in some cases in conjunction with the caloric theory.
Next, Carnot's theorem (e. c.) is verified by a comparison,
necessarily crude, with information derived from experiments
by other authors. Finally, some of the practical consequences
of the preceding analysis for the technology of the steam engine
are pointed out.

I think it is fair to say that Carnot's profound scientific insight was supplemented by luck; Carnot's theorem (e. c.) is an
outstanding case of a true conclusion of capital importance
discovered by correct deduction from premises partly false.
But to balance the picture I must add that Carnot's attitude
toward his false premise, the caloric theory, was anything but
credulous; I shall enlarge on this point below.

     b. Carnot's theorem (e. c.) and the reversibility concept

Pages 9-38 of the *Réflexions* are devoted to the introduction
of the three chief premises above-mentioned, the deduction
from these of Carnot's theorem (e. c.), and the presentation,
in its first and primitive form, of the concept of reversibility.

The argument opens with a brief examination of the steam
engine, which leads to the conclusion that "the production of
motive power in the steam engine is . . . not due to a real
consumption of caloric, *but to its transfer from a hotter to a
colder body* [Carnot's italics], that is to say, to the reestab-
lishment of its equilibrium. " A consideration of the changes
of volume generally associated with changes of temperature
then leads to the generalization: " . . . this statement [that
motive power is due to transfer of caloric] holds not only for
steam-engines but also for all heat-engines. . . ." These
sentences represent an application of the caloric theory to heat
engines and constitute the introduction of premise (2) mentioned above; they also suggest premise (1). There follows the
main theme of the work (already stated in the introductory section), in the form of

. . . the interesting and important question: Is the motive power of
heat invariable in quantity, or does it vary with the agent one uses to
obtain it . . . ? For example, we suppose we have at our disposal

a body, A, maintained at a temperature of 100 degrees, and another body, B, at 0 degrees, and inquire what quantity of motive power will be produced by the transfer of a given quantity of caloric--for example of so much as is necessary to melt a kilogram of ice--from the first of these bodies to the second; we inquire if this quantity of motive power is necessarily limited; if water vapor offers in this respect more or less advantage than the vapor of alcohol or of mercury; than a permanent gas or than any other substance.

Hereupon Carnot explicitly states premise (1): " . . . *wherever there is a difference of temperature the production of motive power is possible* [Carnot's italics]. Conversely, wherever this power can be employed, it is possible to produce a difference of temperature or to destroy the equilibrium of the caloric." The latter sentence contains the first suggestion of the reversibility concept.

Carnot then gives a preliminary argument to answer the "question" just propounded. He considers a process which we should describe as consisting of three of the four steps of a particular Carnot cycle, namely: (1) isothermal conversion of a quantity of water into steam; (2) adiabatic expansion of this steam; (3) isothermal condensation of the product of (2). (Carnot is apparently unaware that step 2 would lead to supercooled steam.) The concept of reversibility is then further developed in the statement: "The operations which we have just described could have been performed in a reverse sense and order, " and this leads immediately to the demonstration which has become classic: "Now if there were any method of using heat preferable to that which we have employed, " then this "method" could be coupled with the reverse of the original operations, and the result would be on the one hand the re-establishment of "things in their original state, " and on the other hand "an indefinite creation of motive power without consumption of caloric or of any other agent whatsoever. Such a creation is entirely contrary to the ideas now accepted, to the laws of mechanics and of sound physics; it is inadmissible. " The last sentence represents the introduction of premise (3) mentioned above. "We may hence conclude that *the maximum motive power resulting from the use of steam is also the maximum motive power which can be obtained by any other means* [Carnot's italics]. " This is the first appearance of Carnot's theorem (e.c.). "We shall soon give a second and more rigorous demonstration of this law." But before doing this, Carnot deals with three other matters.

First and most important, he discusses the new concept of reversibility. He asks, what "in connection with the proposition just stated . . . is the meaning of the word *maximum?* . . . The necessary condition of the maximum is," he concludes, *". . . that in bodies used to obtain the motive power of heat, no change in temperature occurs which is not in turn due to a change of volume* [Carnot's italics]." This condition then requires that transfer of caloric occur, ideally, as in the process above considered, only "between two bodies of equal temperature." But, "In reality things would not occur exactly as we have supposed. In order to effect a transfer of caloric from one body to the other, the first must have the higher temperature; but this difference may be supposed to be as small as we please; we may, in theory, consider it zero without invalidating the argument." Later on (right after the second statement of Carnot's theorem [e.c.]) these conclusions are restated, but Carnot does not go beyond them in the analysis of reversibility. He thus leaves us with his new concept in primitive form.

Second, Carnot points out some respects in which the preliminary argument recounted above is incomplete. And third, he introduces his famous--and in view of the First Law imperfect--analogy of the motive power of heat to that of a waterfall. "The motive power of falling water depends on the quantity of water and on the height of its fall; the motive power of heat depends also on the quantity of caloric employed and on that which might be named . . . *its descent*--that is to say, on the difference of temperature of the bodies between which the exchange of caloric is effected." Then follows the "second and more rigorous demonstration" of Carnot's theorem (e.c.). For this purpose, after a brief description of what we call respectively adiabatic and isothermal changes of volume, the Carnot cycle is introduced, in full-fledged form, for "an elastic fluid--atmospheric air for example--enclosed in a cylindrical vessel." The extension to this case of the demonstration I have referred to above as classic then leads to the famous statement of Carnot's theorem (e.c.): *"The motive pow-*

*er of heat is independent of the agents employed to develop it;
its quantity is determined solely by the temperatures of the
bodies between which, in the final result, the transfer of the
caloric occurs.* " This is followed by the restatement, mentioned above, of Carnot's conclusions regarding reversibility.

c. Consequences for the theory of heat

In pages 38-73 of the *Réflexions* Carnot's theorem (e. c.) is
used as a premise for further deductions, partly alone and
partly in conjunction with the caloric theory as a further premise. This recognition by Carnot of the fertility of his new law
for the theory of heat in general, constitutes, as I have already
said, an important step in the discovery of the Second Law.
His opening sentences reveal his attitude very plainly.

Various methods of developing motive power may be adopted, either
by use of different substances or of the same substance in different
states; for example by the use of a gas at two different densities.
    This remark leads us naturally to the interesting study of aeriform fluids, a study which will conduct us to new results concerning
the motive power of heat, and will give us the means of verifying in
some particular cases the fundamental proposition [i. e. Carnot's
theorem (e. c.)]. . . .

The results in question consist principally of seven theorems
which I have discussed elsewhere[26] under the name of "the
forgotten thermodynamic theorems of Carnot. " They are as
follows:

(1) When a gas passes without change of temperature from one
definite volume and pressure to another, the quantity of caloric absorbed or emitted is always the same, irrespective of the nature of
the gas chosen as the subject of the experiment.

(2) [At a given pressure] the difference between the specific
heat [referred to unit volume] under constant pressure and the specific heat at constant volume is the same for all gases.

(3) When a gas changes in volume without change of temperature the quantities of heat which it absorbs or gives up are in arithmetical progression when the increments or reductions of volume
are in geometric progression.

(4) When the volume of a gas increases in geometrical progression its specific heat [referred to unit weight] increases in arithmetical progression.

(5) [For a given gas at a given temperature] the difference between the specific heat under constant pressure and that at constant

volume is always the same, whatever the density of the gas, provided
the quantity of gas by weight remains the same.

(6) The quantity of heat due to the change in volume of a gas
[at constant temperature] becomes greater as its temperature is
raised.

(7) The descent of caloric produces more motive power at
lower degrees of temperature than at higher.

Since the deduction is in all cases logically impeccable, each
of these theorems must fall into one of the following three
classes: (A) theorems based on Carnot's theorem (e. c.) alone
and therefore true; (B) true theorems based on Carnot's theorem (e. c.) and the caloric theory jointly; (C) false theorems
based on Carnot's theorem (e. c.) and the caloric theory jointly.
A detailed examination, which I omit for lack of space, shows
the actual distribution to be as follows: Class (A), theorems
1, 3; Class (B), theorems 2, 5, 6, 7; Class (C), theorem 4.

d. Experimental verification

Pages 74-89 of the *Réflexions* are devoted to an attempt to
show, by computation from experimental data obtained by other
authors, that "the quantity of motive power produced is really
independent of the agents used. . . ." In the following table I
summarize Carnot's results and compare them with the true
values computed by taking the theoretical efficiency equal to
$(T_1 - T_2)/T_1$, where T_1 and T_2 are the absolute temperatures
of source and sink respectively.

Agent	Number of Units of Motive Power Produced by Descent of 1000 Units of Heat:		
	From 1° C to 0° C	From 78.7° C to 77.7° C	From 100° C to 99° C
Air	1.395	.....	.....
Water vapor	1.290	1.212	1.112
Alcohol vapor	.....	1.230	.....
Any agent, true value	1.556	1.213	1.144

Carnot regards the difference between air and water vapor
at 1° C, and by implication also that between water vapor and

alcohol vapor at 78. 7°, as "not outside the limits of probable error, considering the large number of data of different sorts which we have found it necessary to use . . . [whence] our fundamental law is verified in a particular case. " For completeness I must add that Carnot's unit of motive power is the quantity of work required to lift 1 cubic meter of water through 1 meter (i. e., about 9810 joules) and that his unit of heat is the quantity required to raise the temperature of 1 kg. of water by 1° C (i. e., about 4185 joules).

   e. Practical consequences for the steam engine

Finally in pages 89-118 of his work, Carnot returns to his point of departure, the technology of the steam engine, by calling attention to certain practical considerations arising from his scientific results. None of these considerations is relevant to the inner history of the Second Law except possibly the last, and therefore I mention only this. Carnot estimates-- admittedly very roughly--the motive power obtainable by the combustion of 1 kg. of coal to be 3920 of the units above mentioned and points out that in the best of the contemporary steam engines the efficiency is only about 1/20 of this amount.

   f. Carnot's attitude toward the caloric theory

I have pointed out that of Carnot's three main premises, one, the caloric theory, later turned out to be false. I have also indicated that Carnot's attitude, in the *Réflexions,* toward this premise is critical and prophetic. This judgment is based on three passages which I shall now quote.

The first occurs on page 21, in a footnote attached to the sentence (quoted above in §9b) which introduces the impossibility of perpetual motion as a premise; it reads:

> The objection will perhaps be made that perpetual motion has only
> been demonstrated to be impossible in the case of mechanical ac-
> tions, and that it may not be so when we employ the agency of heat
> or of electricity; but can we conceive of the phenomena of *heat* or of
> electricity as due to any other cause than some *motion* [my italics]
> of bodies, and as such are they not subject to the general laws of
> mechanics ?

The second passage occurs on page 37 in a footnote introduced near the end of the second and rigorous demonstration of Carnot's theorem (e. c.). I quote the entire footnote because the first part establishes beyond question Carnot's employment,

in the *Réflexions*, of the caloric theory as a premise, and the
two final sentences (italicized by me) express his doubts as to
that theory.

We implicitly assume, in our demonstration, that if a body experiences any changes, and returns exactly to its original state, after
a certain number of transformations--that is to say, to its original
state determined by its density, its temperature, and its mode of
aggregation; we assume, I say, that the body contains the same
quantity of heat as it contained at first, or, in other words, that the
quantities of heat absorbed and released in its several transformations exactly compensate one another. This fact has never been
called in question; it was at first admitted without consideration,
and afterwards verified in many cases by experiments with the calorimeter. To deny it would be to overthrow the entire theory of heat,
of which it is the foundation. *It may be remarked, in passing, that
the fundamental principles on which the theory of heat rests should
be given the most careful examination. Several experimental facts
seem to be almost inexplicable in the actual state of that theory.*

The third passage occurs on page 89 in the text proper, at the
end of the section devoted to experimental verification (re-
viewed in §9d above); it reads: "The fundamental law [i. e.
Carnot's theorem (e. c.)] which we wish to confirm seems,
however, to need additional verifications to be put beyond
doubt; it is based upon the theory of heat *as it is at present
established, and, it must be confessed, this does not appear
to us to be a very firm foundation* [my italics]. "

These passages, it seems to me, leave no reasonable doubt
as to Carnot's position in the *Réflexions:* he consciously makes
the best of what he fears is a bad job, by using the caloric theory. Ernst Mach, commenting on the same passages, says:
"S. Carnot stands very close to the kinetic theory of heat, yet
this theory does not break through in his work. "[27] It follows
that to give Carnot the entire credit for the discovery of the
Second Law is an unjustifiable oversimplification. For, his unsurpassed originality and great insight notwithstanding, Carnot's view of the law whose discovery he initiated was both
distorted and incomplete: distorted because he first deduced
it in part from the caloric theory and then used it in conjunction with this theory to deduce further results, one of which
is false (see §9c); and incomplete because, not having the First
Law, he could hardly and certainly did not guess the equiva-

lence of Carnot's theorem (e. c.) to such other generalizations as the Clausius and Kelvin statements (e. c.)--an equivalence which is an essential feature of the Second Law as generally understood.

A just estimate would, I think, emphasize that Carnot deserves particular acclaim for great discovery in spite of the caloric theory. For perspective I may add that Carnot's relation, in the *Réflexions,* to the First Law, is a good deal like that of Vesalius, in the *Fabrica,* to the circulation: Vesalius could not "see . . . how even the smallest particle (of blood) can be transferred from the right to the left ventricle through the septum, "[28] but he did not say it is not transferred; he adumbrated Servetus and Harvey, but did not anticipate them. In the case of Carnot, however, the picture receives its final touch--and here the analogy to Vesalius ends--in the well known fact that in the period of eight years between the appearance of the *Réflexions* and his death, he secretly reached the full truth: he divined the true nature of heat and calculated the mechanical equivalent, probably from the specific heats of gases as was done by Mayer in 1842. These discoveries appear in the posthumous notes already mentioned. [29] There is no sign that Carnot ever combined his new insight with that of the *Réflexions.* This was first done by Clausius in 1850.

g. LaMer's interpretation

The foregoing account of the *Réflexions* is in accord with the almost universally accepted interpretation, which is supported by authorities such as Mach (quoted above). Rather spectacularly divergent from this, however, is another interpretation. Its leading exponents are H. L. Callendar, [30] K. Schreber, [31] J. N. Brönsted, [32] V. K. LaMer, [33] and L. Brillouin. [34] Of these LaMer is the only one who is very active at present, and therefore I shall for brevity speak of *LaMer's interpretation,* as opposed to the *common interpretation.* LaMer's interpretation differs from the common in two respects chiefly.

The first of these hinges upon Carnot's use of the word *"calorique. "* The common interpretation holds that by *calorique* Carnot means heat in its quantitative aspect, i. e., heat measurable in a calorimeter; LaMer's interpretation holds that by *calorique* Carnot means "entropy. " For clarity I must add that

Carnot uses two further words closely related to *calorique*, namely *"feu"* and *"chaleur,"* about the meanings of which La-Mer's interpretation and the common are in agreement: *feu* means fire or any body heated above room temperature by fire, and *chaleur* means heat. Furthermore, in Magie's translation (from which I have taken all my quotations), *calorique* is consistently rendered as "caloric," and *chaleur* and *feu* as "heat." To return to LaMer's interpretation: the principal justification given for translating *calorique* as entropy is that, if this be done, then Carnot's statements come by and large into much better agreement with modern knowledge than they do if *calorique* is translated as heat (since entropy is conserved around a reversible cycle but heat is not).

The second respect in which LaMer's interpretation differs from the common one is in holding that the argument of the *Réflexions* is not based upon the (false) caloric theory at all, but upon the (true) mechanical theory of heat, of which Carnot was, according to LaMer, already thoroughly convinced in 1824. This view is justified principally by appealing to the first of the three statements quoted in §9f above and to the facts revealed by Carnot's posthumous notes. I need hardly say that I think LaMer's interpretation is wrong on both counts, but I cannot do more here than summarize my main objections.

That *calorique* can mean nothing but heat seems to me entirely certain on the basis of the second and third of the statements quoted in §9f. And this is borne out by a number of other passages of which I shall select one of the strongest. On page 24 we read: *"Tout changement de température qui n'est pas dû à un changement de volume ou à une action chimique . . . est nécessairement dû au passage direct du calorique d'un corps plus ou moins échauffé à un corps plus froid."* I submit that *calorique*, if it is capable of "the direct passage from a more or less heated body to a colder body," whatever it is, cannot be entropy: under these conditions heat is *transferred*, but entropy is *generated*, in that the amount received by the colder body is *greater* than that lost by the hotter.

A further objection is the fact that Carnot's false theorem (No. 4 of the seven quoted in §9c above) is deduced by setting equal to zero the sum of the quantities of *calorique* received by a gas going through a certain (not a Carnot) cycle. As for the argument that Carnot's statements become more nearly true if *calorique* is entropy, this is somewhat like saying that Paracelsus, when he maintains that all matter consists of salt, sulphur, and mercury, really means protons, neutrons, and electrons because then he agrees better with what we know today.

Turning now to Carnot's alleged use of the mechanical rather than the caloric theory of heat, this is explicitly denied by the third of the passages quoted in §9f above. Moreover, I fail to see why, if Carnot really did use the mechanical theory, he did not point out that in his cycle the sink receives an amount of heat that is less than that given up by the source, by just the equivalent of the work done in the cycle: of this idea the *Réflexions* shows no trace.[35]

h. The possible influence of L. N. M. Carnot

I have mentioned above that Lazare Carnot, the father of the author of the *Réflexions,* wrote an important book entitled *Principes fondamentaux de l'équilibre et du mouvement.* This work has certain features which in our present context seem striking.[36] Lazare is greatly concerned with the efficiency of machines, and, moreover, not from the traditional Archimedean viewpoint of mechanical advantage, but from that of mechanical energy. Thus, he points out that the theoretical possibility of an infinite mechanical advantage (as in Archimedes' remark about lifting the earth) by no means implies the possibility of perpetual motion. Furthermore, Lazare is at pains to show that because no real bodies are perfectly elastic, therefore, some *vis viva* is always lost in their impact. This leads to the conclusion that "a machine works most advantageously only when in its operation all impact is avoided." Finally, Lazare remarks that "every sort of agency has with respect to its nature or physical constitution a maximum [of efficiency?], which can only be determined by experience." I am unable to read these statements without a certain degree of awe. For we see that the work of Sadi may be described as the inspired extension to thermal engines of Lazare's thinking about mechanical ones. We are reminded again of the Darwins, Erasmus behind Charles.

§10. *From Carnot (1824) to Clausius (1850)*

After the *Réflexions* (1824) the next major step in the dis-

covery of the Second Law was the correlation of Carnot's theorem (e. c.) with the First Law of thermodynamics. This was
accomplished by Clausius in 1850. The intervening period witnessed, as already mentioned, the discovery of the First Law,
and is marked, as far as the Second Law is concerned, by spo-
radic applications of Carnot's theorem (e. c.), usually in conjunction with the caloric theory, as a premise for deduction.
These applications represent a continuation of the type of investigation initiated by Carnot himself in the third part of the
Réflexions, discussed above in §9c. I shall briefly sketch them,
as well as the discovery of the First Law.

   a. Clapeyron (1834)

The first work to make any reference to the ideas of Carnot
was a paper published in 1834 by B. P. E. Clapeyron (1799-
1864), a French engineer, and entitled "Sur la puissance motrice de la chaleur. " Although in this title the sound of the
steam engine is still heard, Clapeyron's express object no
longer includes the improvement of this device but consists
exclusively in the deduction of laws of heat from Carnot's
theorem (e. c.) and the caloric theory. For this purpose he
gives the first graphic representations of Carnot cycles on the
pressure, volume-plane, and points out that the area of the
quadrilateral is equal to the work done. He applies Carnot's
theorem (e. c.) by setting the ratio of the work done by an infinitesimal Carnot cycle to the heat received from the source,
equal to dt/C, where dt is the temperature interval of the cycle
and C is an unknown function of the temperature. This analytical formulation is not original with Clapeyron, inasmuch as
Carnot gives an equivalent one in a long footnote beginning on
page 73 of the *Réflexions,* without, however, using it in the
text. Clapeyron furthermore applies the caloric theory, by
treating heat, Q, as a function of state and hence writing

$$dQ = \left(\frac{\partial Q}{\partial v}\right)_p dv + \left(\frac{\partial Q}{\partial p}\right)_v dp, \tag{1}$$

where v and p denote volume and pressure respectively. From
Carnot's theorem (e. c.) *alone,* Clapeyron deduces, for the
case of a pure liquid in equilibrium with its vapor,

$$\frac{\left(1-\frac{\delta}{\rho}\right)\frac{dp}{dt}}{k} = \frac{1}{C}, \tag{2}$$

where δ is the density of the vapor, ρ that of the liquid, k the heat of vaporization per unit volume of vapor, and the other letters have the meanings already stated. This equation is correct and is the ancestor of the familiar "Clapeyron-Clausius equation" of today; it is Clapeyron's most original result.

From Carnot's theorem (e. c.) and the caloric theory jointly, Clapeyron deduces for any fluid

$$\left(\frac{\partial Q}{\partial v}\right)_p \left(\frac{\partial T}{\partial p}\right)_v - \left(\frac{\partial Q}{\partial p}\right)_v \left(\frac{\partial T}{\partial v}\right)_p = C, \tag{3}$$

and shows that for any gas (treated as perfect) this becomes

$$v\left(\frac{\partial Q}{\partial v}\right)_p - p\left(\frac{\partial Q}{\partial p}\right)_v = RC, \tag{4}$$

where T now denotes temperature and R a constant (equal to the product of the number of moles in the sample by the "gas constant" of the present day, also generally denoted by R). Equations (3) and (4), although as they stand obviously incorrect, are valid physically if $\left(\frac{\partial Q}{\partial v}\right)_p$ is interpreted *merely* as heat absorbed per unit increase of volume at constant pressure and $\left(\frac{\partial Q}{\partial p}\right)_v$ similarly. The equations become entirely correct if we replace $\frac{\partial Q}{\partial}$ by $T\frac{\partial S}{\partial}$, where S and T denote respectively entropy and absolute temperature. From the *incorrect* equations (1) and (3) Clapeyron deduces an equation for the heat dQ absorbed by any fluid in an isothermal change of state which is entirely *correct*, namely:

$$dQ = dv \frac{C}{\left(\frac{\partial T}{\partial p}\right)_v} = -dp \frac{C}{\left(\frac{\partial T}{\partial v}\right)_p} \tag{5}$$

This equation, like equation (2), later played an important role in the work of Clausius (§11). The concluding part of Clapeyron's paper is devoted to the computation, from diverse experimental data, of numerical values of the function C for various temperatures.

Clapeyron's work represents a significant increase in the use of analysis for the deductive exploitation of Carnot's theorem (e. c.). It is also important as the source from which

Thomson and Clausius obtained their first knowledge of the ideas of Carnot (§§6b, 6c).

b. Discovery of the First Law (1839-47)

The credit for the discovery of the First Law belongs chiefly to J. R. Mayer (1814-78), J. P. Joule (1818-89) and H. L. F. Helmholtz (1821-94), but is shared also by M. Séguin (1786-1875) and L. A. Colding (1815- ?). The contributions of Séguin and Colding appeared in 1839 and 1843 respectively. [37] The work of the three chief discoverers is briefly summarized as follows: Mayer in 1842 enunciated the conservation of energy ("Kraft") as an apparent consequence of the philosophical principle of causality, pointed out qualitative empirical evidence for the equivalence of heat and work, and calculated a value of the mechanical equivalent of heat from the heat capacities of gases measured by previous authors. [38] Joule in 1843 published a value of the mechanical equivalent obtained by quantitative experiments on the dissipation of mechanical energy and stated his belief that "The grand agents of nature are, by the Creator's fiat, indestructible. "[39] Helmholtz in 1847 gave the first analytical account (primitive because as yet nonstatistical) of heat as the mechanical energy of the smallest constituents of matter and exhibited the scope and power of the new law by applying it analytically to various fields of physics. [40] It may be added that Mayer published further important communications in 1845[41] and 1848, [42] and Joule in 1845, [43] 1847, [44] and 1850. [45]

c. Holtzmann (1845) and Helmholtz (1847)

Clapeyron's work made plain the importance of the above-mentioned function C (§10a) for deductions drawn from Carnot's theorem (e. c.), and this function plays a large part in subsequent investigations down to 1851. The form of C is given by

$$C = \frac{T}{J},\qquad(6)$$

where T is the absolute gas temperature (or the Kelvin temperature) and J is the mechanical equivalent of heat. The first person to establish this result by an entirely correct method--namely, deduction from Carnot's theorem (e. c.) and the First Law jointly--was Clausius, in 1850. The first person, however, to obtain it was Helmholtz, in his above-mentioned memoir of 1847. Helmholtz's method consists in comparing Clapeyron's equation for gases with an equation deduced by C. Holtzmann in 1845.[46] Holtzmann's work blends elements of the emergent First and Second Laws with the dying caloric theory and illustrates well the sort of creative confusion that frequently precedes a major clarification in science. His equation, which refers to gases, is

$$p\frac{dq}{dp}+\rho\frac{dq}{d\rho}=-\frac{p}{a\rho}\,,\tag{7}$$

where ρ is the density, a is a constant (equal in fact to the mechanical equivalent of heat, although Holtzmann did not grasp the full significance of this), and the derivatives written with d are actually partial derivatives at constant ρ and constant p respectively. This equation is incorrect but physically valid to the same extent as are Clapeyron's equations cited above as equations (3) and (4).

Turning now to Helmholtz, we find a discussion of the work of both Clapeyron and Holtzmann in relation to the conservation of energy. Near the end of this discussion, which is highly incomplete,[47] Helmholtz cites in slightly altered form our equations (4) and (7), and by comparing them deduces

$$\frac{1}{C}=\frac{a}{k(1+\alpha t)}\,,\tag{8}$$

where a is the mechanical equivalent of heat, α is the "coefficient of expansion of gases" (modern value: $1/273.1$), t is the centigrade temperature and k is a constant. This formula, which is equivalent to our equation (6), Helmholtz verifies by comparing C-values computed from it with those found by Clapeyron. Thus we have here a further good example of a true result first discovered by deduction from premises partly false.

  d. William Thomson's discovery of absolute temperature
      (1848)

Of the various consequences known to be deducible from Carnot's theorem (e. c.) without additional premises, the most interesting and important by far is the possibility of defining absolute temperature scales. This deduction was drawn by William Thomson in his first publication bearing on the ideas of Carnot, a short paper of 1848 entitled, "On an Absolute

Thermometric Scale Founded on Carnot's Theory of the Motive Power of Heat, and Calculated from Regnault's Observations. " Thomson begins by pointing out that the thermometry
of his day, although experimentally highly refined, has no
"principle on which to found an absolute thermometric scale, "
inasmuch as it is dependent upon the choice of a particular
thermometric substance, the most favorable of which is air.
He suggests that Carnot's theory furnishes a principle of the
kind in question because "the relation between motive power
and heat, as established by Carnot, is such that *quantities of
heat* and *intervals of temperature* are involved as the sole elements in the expression for the amount of mechanical effect
to be obtained through the agency of heat. . . . ".After some
discussion he gives his definition:

The characteristic property of the scale which I now propose is, that
. . . a unit of heat descending from a body A at the temperature $T°$
of this scale, to a body B at the temperature $(T-1)°$, would give out
the same mechanical effect, whatever be the number T. This may
justly be termed an absolute scale, since its characteristic is quite
independent of the physical properties of any specific substance.

The paper concludes with the mention of two tables[48] for the
conversion of temperature on the air thermometer to temperature on the new scale and vice versa. This conversion requires numerical values of the reciprocal of Clapeyron's function C (§10a), which Thomson denotes by μ, and in his later
papers calls, appropriately, *Carnot's multiplier* or *Carnot's
function*. These values of μ Thomson calculates by applying
Clapeyron's result cited above as equation (2) to the experimental data of Regnault on the vapor pressure and heat of vaporization of water at various temperatures. The scale thus
defined is absolute as Thomson says, but it is not the Kelvin
scale in use today. The latter was defined by Thomson only
in 1854,[49] three years after he had cast off the caloric theory
which in 1848, as is evident from the definition cited above,
was still blinding him (in spite of the publications of Mayer,
Joule, and Helmholtz) to the true relation between heat and
work.

In order to exhibit concisely the logic embodied in Thomson's two absolute scales I shall proceed analytically. Denote
by θ the temperature measured on any scale based on a par-

ticular thermometric substance. Then Carnot's theorem permits us to write, in place of Clapeyron's dt/C (§10a), the expression $\mu(\theta)d\theta$, where $\mu(\theta)$ is *Carnot's function* just mentioned. Carnot's theorem moreover requires that the value of $\mu(\theta)d\theta$ be invariant under all scales θ. From this it follows that if we choose arbitrarily the form of the function $M(T)$ satisfying the equation

$$M(T)dT = \mu(\theta)d\theta, \tag{9}$$

we define a particular *absolute* scale. The choice

$$M(T) = \frac{J}{T} \tag{10}$$

gives the familiar Kelvin scale, and the choice

$$M(T) = \alpha \tag{11}$$

where α is a constant, gives Thomson's first scale, of 1848. For the scale defined by equation (11) it is readily shown that a Carnot engine operating between T and $(T-1)$ on this scale and taking in one unit of heat at T would yield the same amount of work for any value of T, namely: (1) if the caloric theory were true, α units of work (this being the case envisaged by Thomson in 1848); and (2) actually, i.e., under the First Law, $(1-e^{-\alpha})$ units of work. With the help of the First Law it is furthermore readily shown that Thomson's two scales, under appropriate choice of constants, are related by

$$T' = 100\frac{\log T - \log 273}{\log 373 - \log 273}, \tag{12}$$

where T' denotes the temperature on the first scale and T the familiar Kelvin temperature. This formula was appended by Thomson, in a note of 1881, to his paper of 1848 as reprinted in the collected papers.[50] That the T'-scale is highly impractical appears from the fact that at $T=0$, $T'=-\infty$. The purpose of these remarks is to bring out that (1) the existence of absolute temperature follows from Carnot's theorem alone, without additional premise; (2) Thomson's initial choice of scale, though not required by the caloric theory, was suggested by it.

     e. James and William Thomson on the freezing point of
        water (1849-50)

The last of the applications of Carnot's theorem (e.c.) preceding its correlation with the First Law was perhaps the most ingenious. This was the prediction by James Thomson in 1849[51] of the lowering of the freezing point of water by pressure. His paper, which is brief, consists of two parts.

In the first he presents an argument based on Carnot's ideas to show that the effect must occur. This argument I find somewhat difficult to follow, and I furthermore believe it to be fallacious because it assumes equal to zero the work done by a reversible cycle (involving a cylinder filled with air and another cylinder filled with ice and water in equilibrium) that is not isothermal. For clarity I must, however, interject that the lowering of the freezing point by pressure is in truth deducible from Carnot's theorem (e. c.) and the two following facts: (1) ice and water can be in equilibrium over a range of pressures; (2) freezing at constant pressure produces expansion.

In the second part of his paper James Thomson assumes the effect established and computes its magnitude. His method, which is entirely clear and correct, involves consideration of an appropriate Carnot cycle and use of the values of Carnot's function given by William Thomson in the conversion tables computed for his absolute temperature scale of 1848 (§10d). The reasoning is, of course, parallel to that used by Clapeyron in deducing equation (2) above; James Thomson's account, however, shows no appreciation of this fact. His result is

$$t = 0.0075\, n, \tag{13}$$

where t is the lowering of the freezing point in degrees centigrade, and n is the excess of pressure over one atmosphere. This result was excellently confirmed by experiments done by William Thomson and reported in a publication of 1850.[52] This achievement of the Thomson brothers not only constitutes the first really strong confirmation of Carnot's theorem (e. c.), but also ranks high in the history of scientific prediction.

§11. *Clausius' paper of 1850*

To the investigators of the subject between 1845 and 1850, there seemed to be two conflicting theories of the relation between heat and work. There was on the one hand the theory of Carnot, which William Thomson as late as 1849 summarized as follows: "The thermal agency by which mechanical effect may be obtained, is the transference of heat from one body to another at a lower temperature. "[53] On the other hand there was the more recent theory of Mayer, Joule, and Helmholtz, which implied that the production of work by a thermal agency involves the consumption of heat. By 1850 both theories were well supported by experiment: Carnot's theory by the Thomsons' work on freezing point, and that of Mayer, Joule, and Helmholtz by the experiments of Joule. Thomson's awareness of the dilemma appears in a footnote to his paper of 1849, in which he says, in part:

It might appear, that the difficulty would be entirely avoided, by abandoning Carnot's fundamental axiom [i.e., Carnot's theorem (e.c.)]; a view which is strongly urged by Mr. Joule [!] (at the conclusion of his paper "On the Changes of temperature produced by the Rarefaction and Condensation of Air." Phil. Mag. May 1845, Vol. XXVI). If we do so, however, we meet with innumerable other difficulties--insuperable without farther experimental investigation, and an entire reconstruction of the theory of heat from its foundation. It is in reality to experiment that we must look--either for a verification of Carnot's axiom, and an explanation of the difficulty we have been considering; or an entirely new basis of the Theory of Heat.[54]

It was Clausius who first resolved the difficulty--without recourse to experiment--by showing that Carnot's theorem (e.c.) is compatible with the view, derived from the conservation of energy, that in the Carnot cycle *part* of the heat is *transferred* from the high to the low temperature, and the *remainder* is *consumed* in producing the work done. He furthermore brought this synthesis of the two previously conflicting theories into such convincing relation to the available experimental data as to leave no reasonable doubt as to its truth. For these reasons Clausius' paper[55] comes next after the *Réflexions* in importance for the discovery of the Second Law. Since its importance tends to be overlooked at present, I shall quote what Gibbs has to say about it in his obituary of Clausius.

This memoir marks an epoch in the history of physics. If we say, in the words used by Maxwell some years ago, that thermodynamics is "a science with secure foundations, clear definitions and distinct boundaries,"[56] and ask when those foundations were laid, those definitions fixed, and those boundaries traced, there can be but one answer. Certainly not before the publication of that memoir. The materials indeed existed for such a science, as Clausius showed by constructing it from such materials, substantially, as had for years been the common property of physicists. But truth and error were in a confusing state of mixture. Neither in France, nor in Germany, nor in Great Britain, can we find the answer to the question quoted from Regnault. [57] The case was worse than this, for wrong answers were confidently urged by the highest authorities. That question was completely answered, on its theoretical side, in the memoir of Clausius, and the science of thermodynamics came into existence. And as Maxwell said in 1878, so it might have been said at any time since the publication of that memoir, that the foundations of the science were secure, its definitions clear, and its boundaries distinct. [58]

Clausius' paper is entitled "Über die bewegende Kraft der Wärme, und die Gesetze, die sich daraus für die Wärmelehre selbst ableiten lassen. " It is divided into a short introduction and two main parts. The introduction sketches the conflict of theories described above and ends with the promise of its resolution in the sequel. The two main parts are entitled respectively: "I. Consequences of the principle of the equivalence of heat and work, " and "II. Consequences of Carnot's principle in connection with the one already introduced. " I shall briefly review each of these.

Clausius begins Part I by adopting a principle suggested by the general assumption that heat measures the *vis viva* of the particles of matter, namely: "In all cases in which work is produced by the agency of heat, a quantity of heat is consumed which is proportional to the work done; and conversely, by the expenditure of an equal quantity of work an equal quantity of heat is produced. " This is of course the First Law as then newly revealed by Joule and Helmholtz. From this law Clausius deduces, for the first time, the now classic fact that "the total heat of bodies" is not a function of state. He then introduces the equation of state for unit mass of any "permanent gas, " in the form

$$pv = R(a+t), \tag{14}$$

in which, after discussing the deviations recently discovered by Regnault and Magnus, he takes 273 as the best value for the constant a. [59] Hereupon he proceeds to his two chief applications of the equivalence of heat and work, namely, to infinitesimal Carnot cycles undergone respectively by (1) a permanent gas and (2) a liquid in equilibrium with its vapor. For case (1) Clausius obtains the result

$$dQ = dU + A \cdot R \frac{a+t}{v} dv, \tag{15}$$

where U is an undetermined function of v and t, and A is the reciprocal of the mechanical equivalent of heat. For case (2) Clausius' result is

$$\frac{dr}{dt} + c - h = A(s - \sigma)\frac{dp}{dt}, \qquad (16)$$

where

$r =$ heat of evaporation of unit weight of the liquid at temperature t,

$c =$ specific heat of the liquid,

$h =$ specific heat of the vapor remaining saturated while temperature is changed,

$s =$ specific volume of the saturated vapor,

$\sigma =$ specific volume of the liquid.

There follow some preliminary remarks about the value of h (completed in an important way in Part II; see below). The remainder of Part I is devoted to a deductive exploitation of equation (15) with the help of a "subsidiary hypothesis" suggested by the regular behavior of the permanent gases, to the effect that "a permanent gas, when expanded at constant temperature, takes up only so much heat as is consumed in doing external work during the expansion." Equation (15) then yields immediately,

$$dQ = cdt + AR\frac{a+t}{v}dv, \qquad (17)$$

in which c, the specific heat of the gas at constant volume, can be a function of t only. The further results obtained comprise the familiar formulas for the difference between the specific heats at constant p and v respectively, for the relations among p, v, t on adiabatic change,[60] and for the heat absorbed on isothermal change. This last result, Clausius points out, includes that deduced by Carnot (from very different assumptions!), which I have quoted in §9c as "forgotten thermodynamic theorem (3)."

Part II opens with Clausius' main contribution to the discovery of the Second Law: the fusion of the insight of Carnot with that of Mayer, Joule, and Helmholtz. As already said in §8, the crux of this fusion is the deduction of Carnot's theorem (e.c.) from the Clausius statement (e.c.) and the First Law (in Clausius' version, the equivalence of heat and work) as joint premises. Except for the nature of its premises, Clausius' argument follows exactly the lines laid down by Carnot. It consists in showing that two Carnot engines assumed to have unequal efficiencies could by coupling be made to "transfer as much heat as we please from a *cold* to a *hot* body"; this, however, is "not in accord with the other relations of heat, since it always shows a tendency to equalize temperature differences and therefore to pass from *hotter* to *colder* bodies [Clausius' italics]." Thus the Clausius statement (e. c.) is introduced casually, at the end of the deduction. There follows a sentence which marks the first recognition of the fact that *two* laws are necessary as a basis for the theory of heat:

It seems, therefore, to be *theoretically* admissible to retain the first and the really essential part of Carnot's assumptions [here Clausius means that we are to keep Carnot's theorem (e. c.) and reject the idea, based on the caloric theory, that heat is conserved in the Carnot cycle], and to apply it as a second principle in conjunction with the first [i. e., with the principle of the equivalence of heat and work]; and the correctness of this method is, as we shall soon see, established already in many cases by its *consequences*.

We see that although he might well have taken the Clausius statement (e. c.) as the "second principle," Clausius' initial selection for this role is in fact Carnot's theorem (e. c.). It is only in his later papers that Clausius reverses this position.

Having thus clarified the fundamental laws, Clausius turns once more to deduction. He begins by applying Carnot's theorem (e. c.) after the manner of Clapeyron, but without the error caused by treating heat as a function of state. He defines the function of temperature, C, just as does Clapeyron (§10a). He then obtains for a permanent gas undergoing isothermal change

$$\left(\frac{dQ}{dv}\right) = \frac{RC}{v}, \tag{18}$$

and for a liquid in equilibrium with its vapor

$$r = C(s-\sigma)\frac{dp}{dt} \tag{19}$$

Equation (18) follows immediately from equations (5) and (14), and equation (19) is equivalent to equation (2); these relations to Clapeyron's results are pointed out by Clausius. By comparing equation (18), deduced from Carnot's theorem (e.c.), with the alternative expression for the isothermal dQ of a permanent gas deduced from the equivalence of heat and work (obtained by setting $dt = 0$ in equation [17]), namely,

$$\frac{dQ}{dv} = \frac{R \cdot A(a+t)}{v},\tag{20}$$

Clausius finds the very important result

$$C = A(a+t).\tag{21}$$

This result is equivalent to equations (6) and (8), and Clausius is here the first to deduce it correctly. Clausius then compares a series of numerical values of C as given by equation (21) with values obtained by Clapeyron (§10a) and Thomson (§10d) and regards the satisfactory agreement, rightfully, as "powerful confirmation of the two principles and of the subsidiary hypothesis annexed to them." This is the first sharp experimental confirmation of the First and Second Laws jointly. Turning thereupon to the case of the liquid in equilibrium with its vapor, Clausius substitutes equation (21) into equation (19) and obtains

$$r = A(a+t) \cdot (s-\sigma)\frac{dp}{dt},\tag{22}$$

which is the classic result now known as the Clapeyron-Clausius equation.

The remainder of the paper is devoted to applications of this equation to experimental data, chiefly those of Regnault on steam. Of these applications I shall mention three as of particular interest. First, Clausius writes equation (22) in the form

$$Ap(s-\sigma)\frac{a}{a+t} = \frac{ar}{(a+t)^2 \dfrac{1}{p}\dfrac{dp}{dt}},\tag{23}$$

and then calculates each side separately, using Regnault's data, over the range from -15° to 225°C. The excellent agreement obtained constitutes further strong confirmation of the fundamental principles. Clausius furthermore shows that the departure of the values from constancy furnishes a close (though not exact) measure of the departure of saturated steam from equation (14). Second, by eliminating $A(s-\sigma)$ between equations (16) and (22) Clausius obtains

$$\frac{dr}{dt} + c - h = \frac{r}{a+t},\tag{24}$$

and, by inserting into this equation experimental values for r, c, and t, he finds that h, the specific heat of the saturated vapor, is negative. This interesting result had been discovered a few months prior to Clausius' publication by W. J. M. Rankine (1820-72) using another method. Third, by inserting appropriate experimental values into equation (22), Clausius obtains a value of $1/A$, the mechanical equivalent of heat, whose excellent agreement with the values then recently determined by Joule constitutes yet further confirmation of the "first and second principles."

§12. *Thomson's paper of 1851*

Shortly after Clausius had published his discovery of the possibility of reconciling Carnot's theorem (e.c.) with the equivalence of heat and work, William Thomson independently made the same discovery. His paper[61] on the subject appeared about ten months after that of Clausius, and fully acknowledges the latter's priority. The two papers are, however, by no means identical in content, but, on the contrary, Thomson's adds to the developing picture of the Second Law three features so basically important that they must be regarded as part of the discovery of that law (see §8). These features are: (1) the Kelvin statement (e.c.), (2) the deduction of Carnot's theorem (e.c.) from the Kelvin statement (e.c.) and the First Law as joint premises, (3) the recognition of the equivalence, under the First Law, of the Kelvin and Clausius statements (e.c.).

On the other hand, one of the strongest features of Clausius' paper is almost completely lacking from Thomson's, namely, verification of the fundamental laws in question by comparison of their deductive consequences with available experimental data. Thus it was still Clausius' paper that furnished contemporaries with most of the assurance of the truth of the contents of both papers. I might add that Clausius' equations resemble those in use at present more closely than do Thomson's. The chief reason for this is that Clausius evaluates the form of Clapeyron's function C (cf. equations 6, 8, 21), whereas Thomson carries the reciprocal of this function, μ (cf. §10d), unevaluated throughout most of his equations.

Thomson's paper of 1851 is entitled, "On the Dynamical Theory of Heat, with Numerical Results Deduced from Mr.

Joule's Equivalent of a Thermal Unit, and M. Regnault's Observations on Steam. " The paper constitutes the first install-ment of a series which appeared, under this title, at intervals up to 1878. It consists of a short introduction and three main parts. The introduction defines the "dynamical theory of heat" as the hypothesis ("established, " says Thomson, "by Sir Humphry Davy") that "heat consists of a motion excited among the particles of bodies. " This leads to the conclusion, supported by the researches of Mayer and Joule, that "there must be an equivalence between heat and work. " Finally Thomson gives a statement of the objects of his paper which I summarize as follows: (1) to clarify the relation of Carnot's conclusions to the dynamical theory; (2) to compute from Regnault's data on steam the motive power of heat, by means of Carnot's theory in proper conjunction with the dynamical theory; (3) to deduce from Carnot's theory and the dynamical theory jointly, equations of physical interest. These three objects are respectively attained in the three main parts of the paper. Parts II and III consist of thermodynamic deduction of the types already described in §§10 and 11 and, therefore, contain nothing that was any longer both fundamental and new at the time the paper appeared. The great contribution of the paper to the Second Law is all contained in Part I, and, therefore, I shall discuss further only this part.

Part I is entitled "Fundamental Principles in the Theory of the Motive Power of Heat. " Thomson begins by explaining that by a "thermodynamic engine" he will mean an engine working in a cycle between a single "source of heat" and a single "refrigerator" (i. e., what I have called a "simple engine" in §7c). He then states that the theory of heat requires *two* fundamental assumptions, as follows:

The whole theory of the motive power of heat is founded on the two following propositions, due respectively to Joule and to Carnot and Clausius. *Prop. I.* (Joule). --When equal quantities of mechanical effect are produced by any means whatever from purely thermal sources, or lost in purely thermal effects, equal quantities of heat are put out of existence or generated. *Prop. II* (Carnot and Clausius). --If an engine be such that, when it is worked backwards, the physical and mechanical agencies in every part of its motions are all reversed, it produces as much mechanical effect as can be produced by any thermo-dynamic engine, with the same temperatures of source and refrigerator, from a given quantity of heat.

These are Thomson's versions of the First Law and of Carnot's theorem, respectively.

There follows a "demonstration" of Prop. I from the dynamical theory of heat. Up to this point Thomson's standpoint does not differ essentially from that of Clausius. But turning now to Prop. II, he adds something new: "The demonstration of the second proposition is founded on the following axiom:--*It is impossible, by means of inanimate material agency, to derive mechanical effect from any portion of matter by cooling it below the temperature of the coldest of the surrounding objects* [Thomson's italics]. " This is the original version of the Kelvin statement. It is immediately followed by the "demonstration" in question, i.e., by a deduction of Carnot's theorem (e.c.) from the Kelvin statement (e.c.) and the principle of the equivalence of heat and work. This deduction, except for the nature of its premises, follows the lines laid down by Carnot and subsequently used by Clausius as we have seen. Hereupon Thomson emphasizes that Carnot's use of a premise later found to be false (the caloric theory) does not necessarily invalidate his conclusion (Carnot's theorem [e.c.]), and then goes on to say:

It was not until the commencement of the present year that I found the demonstration given above, by which the truth of the proposition [Carnot's theorem (e.c.)] is established upon an axiom [Kelvin statement (e.c.)] which I think will be generally admitted. It is with no wish to claim priority that I make these statements, as the merit of first establishing the proposition upon correct principles is entirely due to Clausius, who published his demonstration of it in the month of May last year. . . . I may be allowed to add that I have given the demonstration exactly as it occurred to me before I knew that Clausius had either enunciated or demonstrated the proposition.

This gracious acknowledgment Thomson then caps with an elegant formulation of the Clausius statement (e.c.):. *"It is impossible for a self-acting machine, unaided by external agency, to convey heat from one body to another at a higher temperature* [Thomson's italics]. " There follows finally the recognition of the equivalence between the two statements: "It is easily shown, that, although this and the axiom I have used are different in form, either is a consequence of the other. The reasoning in each demonstration is strictly analogous to that which Carnot originally gave. "

The remainder of Part I is devoted to some general remarks about the equivalence of heat and work, largely in situations in which electric currents play a role, and, finally, to the deduction from "Propositions I and II" respectively of two equations that are used in the further deductions of Parts II and III, but that we need not consider because they are closely related to equations already obtained by Clausius.

§13. *Thomson's paper of 1852*

About a year after the work just discussed, William Thomson published an inspired paper of three and a half pages entitled "On a Universal Tendency in Nature to the Dissipation of Mechanical Energy. " This paper revealed a new dimension of the Second Law, the cosmological, and it has therefore strongly colored a good deal of general thinking inside and outside of thermodynamics ever since. This revelation, moreover, foreshadowed what I have here called the Generalized Second Law (G-Law, §7b), and I therefore regard it as the last step in the so-called discovery of the Second Law (§8).

After an introductory paragraph Thomson quotes his original version of the Kelvin statement (given in §12), and lays down the following four propositions as

. . . necessary consequences of the axiom. . . . I. When heat is created by a reversible process (so that the mechanical energy thus spent may be restored to its primitive condition), there is also a transference from a cold body to a hot body of a quantity of heat bearing to the quantity created a definite proportion depending on the temperatures of the two bodies. II. When heat is created by any unreversible process (such as friction), there is a *dissipation* of mechanical energy, and a full *restoration* of it to its primitive condition is impossible. III. When heat is diffused by *conduction,* there is a *dissipation* of mechanical energy, and perfect restoration is impossible. IV. When radiant heat or light is absorbed, otherwise than in vegetation or in chemical action, there is a *dissipation* of mechanical energy, and perfect restoration is impossible.

Thomson then illustrates II. by showing that in "the best steam engines . . . at least three-fourths of the work spent in any kind of friction [especially that of the steam rushing through pipes] is utterly wasted"; and he illustrates III. by

deducing an interesting formula for the "work that could be
got by equalizing the temperature of all parts of a solid body
possessing initially a given non-uniform distribution of heat,
if this could be done by means of perfect thermodynamic engines without any conduction of heat. . . ." Finally, Thomson
says:

The following general conclusions are drawn from the propositions
stated above, and known facts with reference to the mechanics of
animal and vegetable bodies:--1. There is at present in the material
world a universal tendency to the dissipation of mechanical energy.
2. Any *restoration* of mechanical energy, without more than an
equivalent of dissipation, is impossible in inanimate material processes, and is probably never effected by means of organized matter,
either endowed with vegetable life or subjected to the will of an animated creature. 3. Within a finite period of time past, the earth
must have been, and within a finite period of time to come the earth
must again be, unfit for the habitation of man as at present constituted, unless operations have been, or are to be performed, which
are impossible under the laws to which the known operations going
on at present in the material world are subject.

For clarity it must be added that Thomson's conclusions
represent not a deduction from the propositions in question,
but rather an inductive generalization. This is in accord with
the fact that the conclusions adumbrate the G-Law, which is
a generalization rather than a deductive consequence of the
usual statements of the Second Law as given in §7c.

§14. *Origin of the term "thermodynamics"*
The term, "thermodynamics, " is due to William Thomson,
and his introduction of it is associated with his work on the
Second Law. His first use of the noun occurs, in the hyphen-
ated form, "thermo-dynamics, " in the second sentence of his
paper of 1854 on thermoelectricity, which constitutes Part
VI of the series entitled "On the Dynamical Theory of Heat. "[62]
Prior to this Thomson had, however, frequently used the ad-
jective, "thermo-dynamic, " in referring to Carnot engines;
the first occurrence is in the paper of 1849 already mentioned
(§§6b, 6c, 10d, 11). I might add that Clausius in his writings
seems never to have adopted the equivalent, *"Thermodynamik";*
he continues to refer to *"mechanische Wärmetheorie. "*

§15. *The aftermath of the discovery*
Of the principal statements of the Second Law given in §7c,

the discovery of that law, which I have just finished recounting, yielded the essential content of only the first three, namely:
   (1) the Clausius statement,
   (2) the Kelvin statement,
   (3) Carnot's theorem.
The emergence of the remaining four kinds of principal statement given in §7c, namely:
   (4) Clausius' theorem,
   (5) the entropy principle,
   (6) the irreversibility statements,
   (7) the Carathéodory statement,
is spread out over the next fifty-seven years after completion of the discovery, i. e., over the period 1852-1909. I shall conclude this article with a relatively brief and incomplete note on each of these four types of statement, and a final note on the G-Law.
   a. Clausius' theorem
The two parts of Clausius' theorem as given in §7c, which refer respectively to reversible and irreversible cycles, I shall call the *equality* and the *inequality*. In 1854 Clausius enunciated both parts, and William Thomson, independently, the equality. The common designation "Clausius' theorem" is therefore unfair to Thomson. The best remedy might consist in calling the entire theorem by some impersonal name, such as the "cycle theorem, " and then, if one wishes, to speak further of the "Clausius-Thomson equality" and the "Clausius inequality. "
   Clausius' paper is entitled "Über eine veränderte Form des zweiten Hauptsatzes der mechanischen Wärmetheorie."[63] Here he defines a very general and abstract concept which he calls the "equivalence-value of a transformation" *(Äquivalenzwert einer Verwandlung)*. This concept plays a basic role in the paper in question and in Clausius' later papers on the Second Law, and Clausius uses it to deduce both the equality and the inequality. The concept is, however, no longer in use today, and for this reason, as well as for brevity, I shall not expound it; interested readers may study Clausius' papers.
   Thomson's paper is Part VI of his series, "On the Dynamical Theory of Heat" and has the subtitle "Thermo-electric Cur-

rents. "[64] The deduction of the equality, which comes near the beginning, uses a method recognizably similar to that commonly employed today, in that the reservoirs of a set at various temperatures, each supplying a quantity of heat to the system undergoing the cycle, are restored to their respective original states by appropriate Carnot engines.

The titles of the two papers just mentioned indicate the diverging attitudes, after 1852, of the two great thermodynamicians toward the Second Law: Clausius set out on a pursuit of general aspects which led finally to his discovery of the entropy, whereas Thomson became increasingly interested in applications to special problems such as that of thermoelectricity. Clausius' title is, to my knowledge, the earliest making explicit reference to the Second Law by that name (the German for "Second Law" is Zweiter Hauptsatz).

  b. The entropy principle

The entropy principle (e.c.) was enunciated by Clausius in 1865. Since this principle is an immediate *mathematical* consequence of Clausius' theorem (e.c.), published as we have seen in 1854, one may well ask why eleven years had to elapse between the two publications. An examination of Clausius' papers suggests the answer that he, at least, was searching for satisfying *physical* interpretations of the line integral along a reversible path:

$$\int \frac{dQ}{T}$$

Already in 1854[65] he had interpreted this integral phenomenologically as a measure of the "quantity of transformation" along the path involved. After 1854 Clausius published four more papers on the Second Law, in 1862, 1863, 1865, and 1867 respectively. In that of 1862[66] he added a *molecular* interpretation of the integral in question, by relating it to two sums over molecular states, which he calls the "heat really present in the body" (actually, the *vis viva* of all the small particles composing it) and the "disgregation" (determined only by the position of the small particles). [67] The paper of 1863[68] is a summary of Clausius' views on the Second Law, at that time. Thus in 1865, when the paper on the entropy appeared, he seems to have felt himself in secure possession of two interpretations, respectively phenomenological and molecular. His molecular interpretation, however, I must for clarity point out, is in no sense an adumbration of the now familiar interpretation of entropy in terms of probability; the latter is due entirely to Boltzmann (1877). Clausius' paper of 1865 is entitled "Über verschiedene für die Anwendung bequeme Formen der Hauptgleichungen der mechanischen Wärmetheorie."[69] The entropy is introduced in the following words (my translation):

If . . . the integral $\int \frac{dQ}{T}$ always receives the value zero whenever the body, after starting from any initial state and traversing an arbitrary set of other states, returns to the initial state, then the expression $\frac{dQ}{T}$ which stands under the integral sign must be the complete differential of a quantity which depends only upon the instantaneous state of the body, and not upon the path by which it has arrived at that state. If we denote this quantity by S, we can set

or . . .

$$dS = \frac{dQ}{T}$$

$$S = S_0 + \int \frac{dQ}{T}.$$

. . . If one seeks for S an appropriate name, he could say--as it is said about the quantity U that it is the *heat-and-work content* of the body--about the quantity S that it is the *transformation content* of the body. But since I think it better to take the names of such quantities important for science from the ancient languages, in order that they might be used unchanged throughout all modern languages, I suggest that the quantity S be called, in accord with the Greek word ἡ τροπή, meaning transformation, the *entropy* of the body. I have intentionally coined the word *entropy* in the closest possible resemblance to the word *energy*, for the two quantities which are to be represented by these words are so closely related in physical meaning, that a certain similarity in their names seems to me expedient.

The conclusion of the paper, which is famous, expresses in terms of the new concept Thomson's cosmological view of the dissipation of mechanical energy:

For the present I shall restrict myself to citing the result that, if one imagines the quantity which for a single body I have called its *entropy*, to be constructed for the entire universe, taking all cir-

cumstances correctly into account, and if one applies also the simpler concept of the *energy*, he can then express the basic laws of the
universe which correspond respectively to the Two Laws of the mechanical theory of heat in the following simple form:
1) the energy of the universe is constant.
2) the entropy of the universe tends towards a maximum.

c. The irreversibility statements

The irreversibility statements (e. c.) of the Second Law are
due to Max Planck (1858-1947). They are foreshadowed in his
doctoral dissertation of 1879, [70] and are given in clear and
complete form in the first edition of his textbook, entitled
Thermodynamik, which appeared in 1897. [71]

d. The Carathéodory statement

In 1909 Constantin Carathéodory (1873-1950) in an article
entitled "Untersuchungen über die Grundlagen der Thermodynamik, "[72] formulated the Second Law in very nearly the
words given in §7c. Carathéodory, at that time active at the
University of Bonn, was a mathematician whose early training had been in engineering. His object in this paper was
to express the Second Law in a form from which the existence of absolute temperature and entropy follows by a method more in accord with the usual mathematical apparatus used
in physics than the argument employing cycles, which is based
on the work of Carnot, Clausius, and Thomson that we have
considered. Carathéodory left untouched the question (answered without proof in §7d) as to how far his statement is
equivalent to the classic ones. His work remained almost un-
noticed until Max Born (1882-   ) called attention to it in
1921. [73]

e. The Generalized Second Law

The history of the G-Law after 1852 (§14) is obscure. The
most concise expression of the law in relatively recent times
that I have been able to find is the following statement by G.
N. Lewis (1875-1946) and M. Randall (1888-1950), in their
text, entitled *Thermodynamics and the Free Energy of Chemical Substances*, which appeared in 1923: "The essential content of the second law might be given by the statement that when
any actual process occurs it is impossible to invent a means
of restoring every system concerned to its original condition.

Therefore, in a technical sense, any actual process is said to be *irreversible.* "[74]*

*I wish to thank Dr. Richard J. Bearman for helping me to clarify my idea of reversibility set forth in §7a, and Miss Ilse M. Bischoff for obtaining for me the portrait of Clausius, published with the permission of the Deutsche Staatsbibliothek zu Berlin, Portraitsammlung. The portrait of Carnot is published with the permission of the Bettmann Archive; the portrait of Thomson is taken from the *The Life of William Thomson, Baron Kelvin of Largs* by S. P. Thompson (London: Macmillan and Co., 1910) and is reproduced with the permission of the Macmillan and Co., Ltd.

PLANCK'S
PHILOSOPHY OF SCIENCE

BY VICTOR F. LENZEN

The dawn of the twentieth century witnessed a revolutionary event in the development of natural science, for on the fourteenth of December, 1900, against the long-standing tradition holding that natural processes are continuous, Max Planck announced the quantum hypothesis which attributes discontinuity to elementary physical processes.

Planck was born in Kiel in 1858, began his scientific studies in Munich, served as professor at the University of Berlin for several decades, and died in Göttingen in 1947. He initiated his scientific career by researches on classical thermodynamics, founded the quantum theory in the course of theoretical investigations on radiant heat, and then enjoyed the good fortune to see the hypothesis of quanta become the basis of all microscopic physical theory. His theoretical physical research was characterized by concern for general principles and unity in physical theory. This philosophic quality of his scientific work was manifested also by reflective analysis of the nature of physical theory and its role in the cognition of reality.

With increasing maturity Planck repeatedly undertook in the course of his long career to expound to the scholarly world the significance of natural science for the culture of his time. The purpose of the present essay is to outline Planck's contributions to fundamental physical theory and to interpret his views on the role of natural science in our *Weltanschauung*.

I

The science of thermodynamics was a relatively new field of physics[1] when Planck began his scientific career. Thermodynamics is the theory of transformations of energy that involve phenomena of heat. Classical thermodynamics is founded upon two general laws which hold for macroscopic, that is, large-scale, processes. The first objectives of Planck's scientific work were the clarification of the laws of thermodynamics and their application to specific problems.

The First Law is founded upon the mechanical theory that heat consists of the disordered motion of the microscopic, that is, fine-scale constituents of physical systems, and expresses the principle of conservation of energy during transformations of systems. By means of the concept of internal energy the principle of the conservation of energy may be formulated as the First Law of thermodynamics in the statement that the internal energy of a system is a function of its state, the increase of which function is equal to the mechanical equivalent of the heat added to the system plus the work done on it. The First Law had been definitely formulated and was generally understood when the fundamental contributions of Clausius to the Second Law aroused Planck's interest in thermodynamics. But the Second Law was still a topic for analysis and provided him with his first problem.

The First Law expresses equivalences during transformations of physical systems, but it does not describe the direction in which they occur. The direction in which natural processes tend to go by themselves is described by the Second Law. The course of development of the Second Law began with a memoir of Sadi Carnot, who analyzed the function of heat engines as they perform mechanical work during cyclic operations. A cyclic heat engine takes in heat from a source at a higher temperature and in the course of the cycle gives up heat to a sink at lower temperature. Carnot attributed the performance of mechanical work in the cycle to the passage of heat from a system at a higher to one at a lower temperature in analogy to the descent of weight from a higher to a lower level. In fact, however, a portion of the heat absorbed at the higher temperature is transformed into mechanical work.

The reconciliation of Carnot's correct conclusions regarding the efficiency of cyclic operations with the principle of conservation of energy was rendered possible by a principle announced by Clausius in 1850. This principle states that heat, by itself, cannot pass from a colder to a hotter body. A more explicit formulation is that heat cannot pass from a colder to a hotter body without compensation. The subject of Planck's doctoral dissertation, *Über den zweiten Hauptsatz der mechanischen Wärmetheorie* (1879), was the clarification and simplification of the principle of Clausius. Planck formulated the principle in the form: The conduction of heat cannot be retraced by any means. He designated a process which cannot be retraced completely by any means as natural, later, as irreversible. In his treatise, *Thermodynamik* (1st ed., 1897), Planck expressed the principle of Clausius by the proposition that it is impossible to construct a motor which in a cycle lifts a weight and only cools a reservoir. This principle signifies that perpetual motion of the second kind, by which work can be derived indefinitely from a source of heat, is impossible.

In accordance with the foregoing definitions, Planck declared that the totality of natural processes falls into two classes, reversible and irreversible processes, according as they can or cannot be completely reversed by any means. Ideal mechanical processes, such as motion in a gravitational field and oscillations of a pendulum, are reversible. The conduction of heat, production of heat by work against friction, and free expansion of gases are examples of irreversible processes which tend by themselves to go in a unique direction. The distinction between reversible and irreversible processes depends only on the constitution of initial state and final state. In a reversible process both states are equally justified. In an irreversible process, the final state is distinguished in a certain sense from the initial state. This distinction was characterized by Planck as a preference on the part of nature for the final state, a measure of which is provided by Clausius' concept of entropy. Planck adopted the principle of Clausius as a self-evident empirical principle and with its aid demonstrated the Second Law of thermodynamics in the form that in every natural process which involves a number of bodies, the sum of the entropies of all participating bodies increases; in the limiting case of reversible processes the total entropy remains unchanged.

In his reminiscences, Planck reported that his discussion of the Second Law of thermodynamics was largely ignored. Kirchhoff rejected his conclusions with the remark that entropy, the magnitude of which is measurable only by a reversible process, cannot be applied to irreversible processes. Planck, however, deemed entropy to be the most important property of a system beside energy. The maximum of entropy determines equilibrium; hence, knowledge thereof yields the laws of physical and chemical equilibrium.

Planck then turned his attention to the First Law of thermodynamics in response to a competition which was set by the philosophical faculty of the University of Göttingen. The result was a work, *Das Prinzip der Erhaltung der Energie* (1887), for which he received a second prize, the only one awarded by the faculty. In this work were described the historical development of the principle, the formulation and proof thereof, and the different kinds of energy. The principle of conservation of energy was held to be founded upon the empirical generalization that perpetual motion of the first kind, by which work can be obtained indefinitely from nothing, is impossible to realize.

Planck further applied the methods of thermodynamics to changes of states of aggregation, to mixtures of gases, and to solutions. He obtained fruitful results but discovered later that his researches had been anticipated by J. W. Gibbs.

II

As we have seen, Planck's initial researches were in the field of thermodynamics, which describes transformations of general properties of macroscopic systems. During the last decade of the nineteenth century he undertook the study of microscopic systems in order to create a theory of the partition of energy within radiant heat. Kirchhoff had initiated the subject by introducing the concept of normal radiation as a function of temperature only. Radiation, within a cavity bounded by the totally reflecting walls of a material enclosure, is normal when it is in equilibrium with emitting and absorbing matter within the enclosure. Since the specific properties of an emitting body do not enter into the constitution of normal radiation, Planck assumed that the material agent of emission and absorption consisted of two oppositely electrified particles which vibrate with respect to each other. He attacked the problem successively by classical mechanics and electrodynamics, by thermodynamics, and by atomic theory.

By classical dynamics and electrodynamics, Planck investigated the exchange of energy between an oscillator and a field of radiation and obtained the energy of radiation for a given frequency as a function of the energy of the oscillator. Since the result was quite general, he then directed his attention to the energy of the oscillator.

Planck next pursued the problem by the methods of thermodynamics. Unlike other investigators, he studied the properties of an oscillator from the standpoint of entropy and its dependence upon energy. The reciprocal of the second derivative of entropy with respect to energy, which may be designated by R, played a basic role in the discussion. The physicist Wien had discovered a law of radiation that was confirmed experimentally for short wave lengths; from this result Planck found that R was proportional to the energy. Experimental results for long wave lengths, however, required that R be proportional to the square of the energy. He then obtained the correct radiation law by expressing R as the sum of two terms, respectively proportional to the energy and to the square of the energy. The law of radiation determines that the intensity of radiation exhibits a maximum for a specific wave length which depends on temperature.

Planck then undertook to provide a theoretical foundation for his law; to accomplish this purpose he abandoned the general method of thermodynamics and utilized atomistic concepts that had been developed by Boltzmann. In order to obtain the required expression for the energy of an oscillator, Planck made the hypothesis of quanta. The quantum hypothesis was that an oscillator in a field of radiation emits and absorbs energy discontinuously by integral multiples of a finite quantum of energy which is equal to the product of a constant, the elementary quantum of action, and the frequency of vibration of

the oscillator. The atomistic method was derived from Boltzmann, who had given a statistical interpretation of the Second
Law of thermodynamics, based on the characterization of
states of systems in terms of probability.

Boltzmann had applied his method to a gas which on kinetic
theory is constituted by many molecules. The macroscopic
properties of a gas, such as temperature and density, are determined by the average values of quantities that characterize
many molecules. These average values are determined by
laws of distribution, such as that of energy among the molecules. If the range of a molecular quantity is subdivided into
cells, an individual molecule may be described in terms of a
value of that quantity by assigning the representative point of
the molecule to its appropriate cell. A particular assignment
of all molecules to the various cells constitutes a microscopic
state of the system. The macroscopic states of a gas depend
only upon the numbers of molecules distributed among the accessible cells. On account of similarity, molecules may be in-
terchanged between cells without change of macroscopic state.
Thus, each macroscopic state of the gas can be realized by a
multitude of microscopic states. The number of microscopic
states that realize a given macroscopic state determines the
probability of the macroscopic state. The entropy of a system
in a given macroscopic state is proportional to the logarithm
of the probability of that state. Boltzmann's statistical interpretation of the Second Law of thermodynamics was that
if a system is left to itself it probably will pass from a state
of less probability to one of greater probability. Equilibrium
is the state of maximum probability.

Planck applied the statistical method of Boltzmann to calculate the probability of the state of an oscillator with given energy. Now, the distribution of energy over a single oscillator
in time could be represented by the instantaneous distribution
of energy over a large set of oscillators. The total energy of
the set of oscillators was subdivided into an integral number
of finite quanta of energy. The macroscopic state of the set
was described in terms of the distribution of quanta of energy
among the oscillators. It was then possible to calculate the
probability of the state and, further, the entropy of the set
of oscillators. The distribution for the state of equilibrium

was found and, hence, the average energy of an individual oscillator as a function of frequency and temperature. From the dependence of energy of radiation upon energy of oscillator, Planck obtained the law of normal radiation. He announced his epoch-making result at a meeting of the Physical Society in Berlin, on December 14, 1900. Thus originated the quantum theory that has introduced discontinuity into elementary natural processes. For the example of the oscillator, Planck had discovered the discrete stationary states of an elementary physical system. His contributions to the theory of heat radiation were set forth in *Theorie der Wärmestrahlung* (1st ed., 1906).

In his initial version of the quantum theory, Planck assumed that an oscillator emitted and absorbed energy in discrete amounts that were integral multiples of a quantum. He sought, however, to minimize quantal properties as much as possible and developed a revised theory, according to which absorption occurs continuously but emission discontinuously. Ultimately, the factor of discontinuity became dominant. Einstein introduced the hypothesis that radiation itself is constituted of quanta, or photons; Bohr discovered the discrete stationary states of the hydrogen atom. Thus, the quantum of action came to be accepted as the basic characteristic of processes of elements of matter and radiation. The originator of the quantum theory lived to see the creation of a system of quantum mechanics as the basis of atomic physics.

III

During Planck's scientific career there was much discussion of the status in reality of the objects of scientific theory. [2] Interest in this problem was stimulated in part by the philosophical criticisms of Ernst Mach, who developed his views in opposition to the theory of knowledge generally held in scientific circles. Upon the development of exact science in the early modern period, a theory of cognition came to be accepted to the effect that the object of perception is independently real and not directly known by the observer. The real object was held to be perceived through the intermediary of sensory data which it produces. The dualist theory of an independent reality behind the data of perception was held by Galileo and Newton and was criticized by empiricist philosophers, such as Berkeley and Hume.

During the nineteenth century, criticism of dualism became a constituent of scientific thinking through the contributions of the physicist, Mach. He viewed the dualism between independent object and datum of perception as metaphysics that is to be eliminated from scientific thought. Mach described the object of perception as constituted of neutral elements that are intrinsically neither physical nor mental, but acquire such status by their relations. Elements of sensation were deemed to be physical in relation to one another; they were deemed to be mental in relation to the bodily organism, itself a complex of elements. During the twentieth century, the doctrine of Mach became a constituent of the logical positivism of the *Wiener Kreis*.

Planck repeatedly expressed his criticism of Mach's theory. As is indicated by the title of an address, "Positivismus und reale Aussenwelt" (1930), he designated Mach's theory of knowledge by the term positivism although an alternative term was phenomenalism. As a basis for his analysis of cognition, Planck acknowledged that the source of every cognition, and hence, the origin of every science, is constituted of personal experiences. The material of every science is received directly through personal perception or indirectly through reports of perceptions by others. These immediately given, personal experiences are elements of indubitable reality and provide the basis for those sequences of thought that constitute science. Physical science is especially characterized by descriptions of objects in terms of the results of measurements also founded upon perception.

Having acknowledged the basic role of perception in science, Planck then asked whether positivism is justified in the claim that such a foundation alone is adequate for physical science. He conceded that positivism does not exhibit logical contradiction, for two experiences cannot contradict one another. Inasmuch as positivism restricts its concern to problems that can be answered by perception, its pronouncements are characterized by illumination and clarity. In support of his rejection of positivism, Planck declared that it is not entirely simple to apply that doctrine in every case. We continually depart from it in the speech of daily life. According to positivism, a table is only a complex of sensations that ceases to exist when perception ceases, but in daily life we view the table as something more than the content of perception. The same consideration holds similarly for plants, animals, and human beings. On positivist theory, it is incorrect to speak of deceptions of the senses; we are deceived only by erroneous inferences concerning them. Interpretations of an optical phenomenon, for example, the bent appearance of a straight stick that is partly immersed in water, and interpretations of astronomical observations by means of a Ptolemaic or Copernican theory are, for positivism, free inventions of the mind that are determined by considerations of utility.

Science requires the reports of a number of investigators. Now, one must acknowledge the distinction between personal experiences and experiences of other persons, which can only be inferred indirectly and, hence, should be viewed by positivism as useful fictions. Planck declared that for positivism an individual scientist is not justified in utilizing the reports of others. He concluded that the scientist must either abandon the ideal of a comprehensive science or compromise the positivist standpoint and utilize the observations of other scientists. But modification of positivism to the extent of employing the experiences of others is not yet a sufficient basis for science. If reports of observations are accepted merely as evidence of experiences, all honorable and trustworthy physicists in the past and present would have the right to claim authority for their own experiences. It would then not be understandable that reports of past scientific discoveries by reputable investigators are completely ignored. One would have to add that only data reported by a few experimental physicists with especially sensitive instruments are of value. Planck further argued that it is difficult to comprehend from the positivist standpoint that individual experiences, such as those by which electromagnetism, electromagnetic induction, and electric waves were discovered, should have produced such revolutionary results in the world of physicists. The necessary ground of this striking phenomenon is an objective physical world independent of the investigator. The positivist denies this presupposition because he acknowledges no reality other than the experiences of the individual physicist.

Planck conceded that positivism has a firm foundation but maintained that it is too restricted. A supplement is needed to free science from the contingencies introduced by human individuals. He held that this is achieved by a step into metaphysics, which is required by healthy reason, through the hypothesis that immediate experiences do not themselves constitute physical reality, but that behind immediate experience there exists a real, external world of which our perceptions give notice. From this point of view the so-called useful inventions of positivism acquire a higher grade of reality than the immediate data of sense. An element of the irrational is thereby introduced into science. The work of science is pursuit of a goal that never will be attained and in principle never can be attained. For the goal is a metaphysical one that lies behind every experience.

Planck declared that the goal of the physicist is cognition of the real, external world. He thus adhered to the view of Helmholtz that experiences are not pictures of the real world but signs thereof from which one draws conclusions. The only means of investigation are measurements which do not convey direct knowledge of the real world. The physicist must, and does, presuppose that a reasonable significance resides within the foundations. He must presuppose that the real world conforms to certain and incomprehensible laws, even though he has no prospect of grasping these laws completely. In reliance upon the regularity of the real world, he forms a system of concepts and propositions that constitutes his physical world picture; it is so fashioned that when substituted for the real world it sends him the same messages as does the latter. In so far as this project succeeds, the physicist may assert, without fear of refutation, that he has actually cognized an aspect of the real world, although this assertion cannot be directly verified. As is indicated by the title of his first general address, "Die Einheit des physikalischen Weltbildes" (1908), Planck especially sought to create a unitary world picture. To this end he bestowed particular attention upon the principle of least action as a principle that regulates all reversible processes.

Planck's theory of knowledge thus differentiated the realm of sensory experiences, the physical picture of the world, and the real world. From the positivist standpoint, the function of the world picture is to describe the sensory world as simply as possible; from the metaphysical standpoint, its function is to cognize the real world as completely as possible. Planck adopted the metaphysical point of view of dualism and declared that the task of the physical picture is to establish the closest possible connection between the real world and that of experiences. Sensory experiences provide the material processed so as to separate and eliminate, as far as possible, everything characteristic of human sense organs or measuring instruments. The elaboration of the physical world picture involves departure from the sensory world and signifies progressive approach to the real world. In the address cited above, he identified the real world and the limit of the world picture. This Kantian point of view may have been a relic of an earlier critical view that he did not express in later discussions. The physical picture of the world has only one condition to satisfy, that of freedom from contradiction; the theorist may otherwise exercise freedom of construction. The physical world picture is constituted of hypotheses, each of which is a product of the free speculations of the human spirit. The usefulness of an hypothesis is to be tested only by the consequences to be derived therefrom. From another point of view, it is the theories constituting the world picture which interpret physical measurements and thereby give them meaning. Planck's theory of knowledge is summarized by the statement that the physical world is only a self-created picture of the real world and, therefore, complete cognition of the physical world represents self-understanding.

IV

Einstein once drew a distinction between physical theories of principles and constructive physical theories. As we have seen, Planck contributed to both kinds of theory: to thermodynamics, a theory of principle that rests on broad generalizations from experience; to quantum theory, a constructive theory for the elements of matter and radiation. The two types of theory describe physical phenomena in terms of different kinds of physical law, namely, dynamical and statistical regularities. It accorded with Planck's concern with foundations that he directed his analysis to the concept of causality in the light of dynamical and statistical laws.

Causality is exemplified by the processes of ideal dynamical systems, such as the vibrations of an ideal pendulum in conformity to an invariable law. In a dynamical process an earlier event, the cause, is necessarily succeeded by a later event, the effect. Planck employed the term dynamical to designate laws that express regularity of connection. All reversible physical processes conform to dynamical laws. In classical physics the actions of atomic constituents of physical reality also were assumed to satisfy dynamical laws. Observable phenomena of physical systems were then explained as constituted by the action of a large number of elementary processes which elude individual description. The resultants of the elementary processes fluctuate slightly about an average and, hence, laws for the macroscopic manifestations of atomistic processes express statistical regularities. Thus, the Second Law of thermodynamics, according to which heat by itself flows only from a higher to a lower temperature, states that the conduction of heat, on the average, proceeds irreversibly in a unique direction. It is not impossible, though very improbable, that heat will flow in the reverse direction without compensation in the environment. Dynamical laws express certainties that are founded on necessities; statistical laws express expectations that are founded on probabilities. Throughout his career in maturity, Planck studied the relation between dynamical and statistical regularity, between causality and probability.

In an address, "Dynamische und statistische Gesetzmässigkeit" (1914), Planck set forth the classical view that statistical law is based upon dynamical law, that probability is founded upon causality. He held that the exact calculation of probabilities requires the persistence of dynamical laws for microscopic elementary processes. A dynamical law satisfies the demand for causality; a statistical law, however, presents the problem of analysis into dynamical elements. At the time when Planck set forth this view, a quantum mechanics for the

processes of the elements of matter and radiation had not yet
been created. He viewed dynamical and causal law to be fundamental; statistical law, based upon probability, was held to
be derivative from the former.

Planck also concerned himself with the philosophical problem of the nature of causality. His theory of causality progressed to the stage that he acknowledged predictability to be
the criterion of causality. Dynamical laws express causality
in that they make possible prediction of individual events; statistical laws make possible the prediction of average values.

Although Planck originated the quantum theory, it remained
for others to create a quantum mechanics for the microscopic
constituents of physical reality. A central constituent of the
theory is representation of the state of an atomic system by a
wave function which has been interpreted as a probability amplitude. The wave function which represents the state of a system determines the probabilities that specific values of characteristic quantities of the system will be found upon observation. Quantum mechanics thus introduced probability and statistics into the foundations of microscopic physical theory.
This elevation of statistical regularity to a fundamental position did not win the approval of the discoverer of the quantum
theory of action. To the problem of necessity versus contin-
gency, causality versus probability, determinism versus indeterminism, he devoted a number of addresses.

In an address, "Die Kausalität in der Natur" (1932), Planck
reaffirmed that predictability with certainty is the criterion
of causality. But he conceded that this criterion does not express the full sense of causality. We can predict that night
will succeed day, but the relation between them is not causal.
Conversely, events that are unpredictable may be deemed to
be the effects of causes. The weather, for example, is assumed to be the effect of mechanical, thermal processes but
cannot be predicted with certainty. The attempt to improve
reliability of prediction leads to the investigation of a limited
portion of the atmosphere, but even then absolute certainty
eludes us. Planck further stressed the fact that since measurements lack precision, it is not possible to predict the result of any measurement with certainty. Since he had adopted
predictability as the criterion of causality, Planck found him-

self before a dilemma: either to abandon causality in nature or to modify the criterion in order to provide for causality.

He acknowledged that many contemporary physicists have accepted the first alternative, and hold indeterminism to be fundamental in nature. On this view the basic laws of physics are statistical laws that express probabilities of results of observation. Planck, however, declared that the second alternative conforms to the traditional outlook of physical science. In classical physics an event of sensory experience is represented for theory by an event in a conceptual world, the world picture of physics, in which strict causality holds between events as conceptual objects. It was hoped by classical physicists that progressive refinement of measurements would reduce the discrepancy between sensory and conceptual event. Heisenberg's principle of indeterminacy, however, shows that the precision of correlation of classical concepts to sensory objects is limited by the finiteness of the quantum of action, so that precise measurement of quantities of position and time destroys the precision of values of momentum and energy respectively. Planck then pointed out that representation of a system by a wave function which satisfies a dynamical law restores causal description. Thus, determinism holds in the physical picture that is based upon quantum mechanics.

Planck further presented a method for the preservation of causality which is exemplified in philosophical absolute idealism. Cognition is a relation between subject and object. In the preceding discussion, the conceptual object in the physical picture was modified so as to restore causality. But Planck declared that one can preserve causality also by modification of the subject. He conceived of an ideal subject that can directly apprehend the sensory world without the mediation of sense organs or instruments. To such an ideal intellect the natural world would exemplify causality. In answer to criticism that there is no empirical foundation for such an ideal intellect, Planck cited the great investigators who were inspired by faith in a rational world order. The very possibility of comprehending and controlling nature suggests that there exists a harmony between the external world and human intelligence. The most complete harmony and therewith the strictest causality reaches its pinnacle in the assumption of an ideal spirit which intuits the entire panorama of events.

Planck's final statement on the law of causality was that it is neither true nor false, but is a heuristic principle which serves to guide physical science in the task of representing the realm of phenomena as an orderly system.

The appeal to an ideal intellect, which transcends the human intellect and cognizes a causal order without disturbing constituent events, points the way to a solution of the apparent contradiction between the application of causality to life and mind, on the one hand, and the experience of freedom of the will, on the other. Planck resolved the contradiction by a distinction between external and internal points of view. From the internal point of view, I experience myself as a being who is free to act. Self-observation disturbs volition and, hence, does not meet the requirement of scientific objectivity. Observation from an external point of view, which is exemplified by recollection of one's own past mental states, will find that human action is determined by motives and conditions in accordance with the law of causality. For Planck, the contradiction between personal experience of freedom and the applicability of causality thereto is an example of a *Scheinproblem*.

V

It accorded with the breadth and depth of Planck's philosophic outlook that he should express himself concerning the role of religion in the present scientific age. He did discuss this subject in one of his last addresses, "Religion und Naturwissenschaft" (1937). The subject was introduced by quoting the answer of Faust to Gretchen's anxious inquiry as to his stand on religion, "Don't want to rob anyone of his sensitiveness and his church." Planck did declare that accounts of miracles which violate natural laws cannot be accepted in this scientific era. On the other hand, he vigorously rejected the movement of godlessness.

According to Planck, the foundation of religion is belief in a supernatural power whose omnipotence and benevolence gives man comfort and peace of mind in face of uncertainties and perils in life. It is characteristic of all religions that this su-

pernatural power is given a personal, or at least anthropo-
morphic, form. The practice of religion involves creation
of symbols that assume different forms. Planck acknowledged
that religious symbols and corresponding rituals are necessary, but he deplored conflicts that have occurred in the past
over divergent doctrines. The fundamental faith of the religious person is that behind all symbols stands a God who as-
sures man that the world is a rational order in which his purposes can be fulfilled. This religious faith is supported by
the fact that natural science finds that an independent nature,
whose inner essence is unknown, is ruled by laws which can
be formulated so as to exhibit purposive, rational activity.

In support of his thesis that the doctrines of religion accord
with the foundations of science, Planck restated his rejection
of positivism in the theory of knowledge. He interpreted positivism as a doctrine of criticism that would limit science to
measurements. He asserted, however, that positivism requires the presupposition that the result of every measurement can be reproduced. A result should depend neither upon
the time and place of measurement nor upon the personality
of the observer. Planck contended that this objectivity of results of measurement is founded on a real world independent
of the observer. Thus, science does require the existence of
a world beyond sensory experiences. Science further has revealed that man is only a tiny creature on a small planet in a
vast cosmos. The faith on which religion is based is matched
by the faith in a universal and objective order as the foundation of science. Planck thus held that the unity and order religion ascribes to the universe has its parallel in the unity
and order embodied in the laws of the physical world.

In the effort to create a unitary foundation for physical science, Planck had long concerned himself with the principle
of least action. He found a model for this principle in the doctrine of Leibniz that God created the actual world with its
distribution of good and evil so that it is the best of all possible worlds. Leibniz formulated the concept of action, and
Maupertuis introduced the principle in mechanics that motion is determined by God so that there occurs the least expenditure of action. An adequate form of the principle of least
action is Hamilton's principle that the motion of a dynamical

system from a fixed initial configuration to a fixed terminal configuration during a prescribed time occurs so that the integral of the difference of kinetic and potential energy has an extreme value when compared with the values for an infinitely near, varied motion. The principle thus describes motion through the proposition that a definite integral has an extreme value, which under certain conditions is a minimum for the natural motion as compared with possible neighboring motions. The differential equations of motion can be derived as the necessary conditions for the validity of Hamilton's principle of stationary action, a variational principle that expresses the laws of motion by means of an integral.

The principle of least action was formulated for mechanics. Helmholtz and his successors have found appropriate functions for electrodynamics, thermodynamics, and the newer theories of relativity and quanta, so that the various fields of physics can be based upon a variational principle formulated in terms of an integral. The integral principle and the differential equations which constitute its necessary condition, are equivalent modes of description which differ only in mathematical form. In the opinion of the writer no philosophical consequences are to be drawn from the formulation of physical laws in terms of a principle of least action. In his discussions of theoretical physics, this also appears to have been the view of Planck. But in the previously cited address on religion he appears to have taken a different position. In this address the principle of least action was interpreted to describe a natural process as directed to the end that a certain integral has a least value. A motion that occurs in nature is selected out of all motions so that the goal is most economically attained. Planck interpreted this mode of determination as one by final cause, by teleology. In his view, the principle of least action describes nature as ruled by a purposive, rational will.

Further scientific support for religious doctrines may be obtained from Planck's previously cited provision for causality in the real world. As we have seen, he had recourse to an ideal intellect by which exact causality is known to exist in the world, and this also constitutes an argument for a conception of God. The idealist philosopher Royce, for example, argued that the distinction between truth and error for

a finite knower requires the existence of an absolute mind
by which truth or error of ideas is directly known.

Planck thus concluded that science and religion are in harmony. Both modes of thought presuppose that there is a rational world order which is independent of man and cannot be
directly known. The methods of science and religion do differ.
Science cognizes nature through measurements; religion employs characteristic symbols. The instinctive intellectual demand for a unified picture of the world requires, however,
that the order of nature be identified with the order of God.
Natural science and religion thus complement one another. To
natural science God is the goal; to religion he is the starting
point of all activity. Planck concluded his unifying discussion
of science and religion with the declaration that together they
are fighting a battle against skepticism and dogmatism. [3]

THE DEVELOPMENT
OF ETHNOGRAPHY
AS A SCIENCE

BY ROBERT H. LOWIE

Like other sciences, ethnography has a long prescientific past. Even savages could not help noting that alien groups had tools, weapons, and customs different from their own. Australian aborigines, obliged to get their diorite for ax-heads from distant parts of the continent, met fellow-blacks with strange ways of life, some of which they rejected while eagerly adding others to their own inventory. On a higher level, Herodotus (?490-24 B. C.) not only noted what he saw in Egypt and Babylonia, but systematically recorded his observations, thus becoming the father of ethnography no less than of history. Comparable information was gathered outside of occidental civilizations. The Chinese General Chang K'ien, who set out on a mission to the west in 138 B. C., returned in 126 B. C. and "submitted to his astounded countrymen a glowing account of the new world which he had discovered, and which was nothing less than the Hellenic-Iranian civilization inaugurated in those regions by the successors of Alexander the Great."[1] Chinese scholars generally registered the customs of their neighbors. For example, they tell us that in 307 B. C. Wu-ling, king of Chao, adopted the cavalry tactics of the northern nomads; and that even the ruler of Tibet abandoned felt dress only as late as A. D. 641.[2]

The Middle Ages must be credited with outstanding accessions to ethnographic knowledge. To cite only two illustrious names, there were the narratives of Ibn Batuta (1304-77) and of Wilhelm Rubruk or Ruysbroeck (ca. 1210-70). The Arab traveler describes, among other things, the shocking liberty enjoyed by women of the Tuareg people in the Sahara; the Fleming graphically pictures the economic life, divinatory practices, and what not of the Mongols.

Excellent as were many of these early accounts, they represented only raw material for a prospective science so long as the majority of human societies remained literally unknown and the known ones accordingly could not be seen in perspective. A zoologist would have a curious conception of the animal kingdom if he knew only about mammals and arthropods. In this respect the mere discovery of new territories, for instance, of Australia in 1605, was of no direct moment. But in America the first contact with Indians was soon followed by full reports on their arts, manners, and beliefs--witness Fray Bernardino de Sahagun's exhaustive account of Mexican religion, based on inquiries probably started about 1530. In the South Pacific, Captain Cook, himself a good observer, took with him such scientifically trained men as George Forster, Johann Reinhold Forster, Sir Joseph Banks, and Dr. Solander, the "ingenious and learned Swede" and "disciple of Linnaeus." From the new data on American, Polynesian, and Melanesian natives provided by naturalists, missionaries, and traders, a respectable body of knowledge for the major divisions of the globe had accrued before the middle of the last century. In 1843 Gustav Klemm thus ventured to publish his *Allgemeine Cultur-Geschichte der Menschheit,* soon to be supplemented by a *soi-disant* treatise on *Allgemeine Culturwissenschaft.*

Educated reporters automatically went beyond their descriptive findings, for these recalled things they had read or themselves seen elsewhere. The Jesuit Joseph François Lafitau (1670-1740) has been, with considerable justice, pronounced an outstanding precursor of ethnographic *science.* Having spent five years in eastern Canada and profited from the more prolonged experiences of a fellow-missionary, he did not content himself with describing what he had learned but systematically looked for parallels in classical antiquity, trying to illuminate Indian by Greek or Roman customs, and vice versa. That today his comparisons often strike us as odd is not to be wondered at, but Lafitau's *Mœurs des sauvages américains, comparées aux mœurs des premiers temps* (Paris, 1724) cer-

tainly does not merit the depreciatory comment by a recent
historian, who regards the work as merely "a very generalized
rehash of the *Jesuit Relations.*"[3]

In the sequel all sorts of wild theories developed from anal-
ogies stressed by otherwise meritorious observers. Suffice
it to mention but two. According to George Catlin (1796-1872),
the Mandan Indians of the upper Missouri were descended from
Welsh immigrants because they crossed rivers in hide-covered
tublike boats ("bull-boats") resembling the Welsh coracles.
Again, James Adair in *The History of the American Indians*
(London, 1775) identified the New World aborigines with the
Ten Lost Tribes of Israel because Jews and Indians shared
menstrual and dietary taboos. On this theory Swanton has offered the obvious comment: "Taboos were so numerous with
the old time Indians that parallels with the taboos of any other
nation would be found without a great deal of difficulty."[4] As
for Catlin's conclusion, bull-boats would prove Tibetan as
well as Mandan affinity with the Welsh. Ignorance of geographical distribution underlay many fallacious conclusions, then
and later.

By no means all of the earlier suggestions, however, be-
long to the realm of fancy. Scientists, automatically applying
their accustomed professional procedures, collated their own
findings with their predecessors', whether in the same tribe
or elsewhere, and often drew sound conclusions or broached
basic problems. Thus, George Forster inquired into the effect
of climate on a people's manners and ethos; further, human
irrationality struck him forcibly when he beheld nude Yahgan
Indians ornamenting their shivering bodies. The shortcomings
of man's intellect also impressed Alexander von Humboldt when
he contrasted the restricted range of the pre-Columbian potato
with the early and wide diffusion of tobacco. Racial psychology
engaged the attention of Humboldt and Prince Maximilian of
Wied-Neuwied, both of whom conceived a favorable opinion of
aboriginal American intelligence. Having heard a young Carib
preach, Humboldt concluded that the orator's stock was capable
of high cultural development *(ein begabtes, einer hohen Cul-
turentwicklung fähiges Volk).*[5]

Nevertheless, even the best-trained scientists of the earlier
periods could not fully attain the ends of an ethnographic *sci-*

ence for the simple reason that the central aim of such a science had not yet crystallized. As Kroeber and Kluckhohn set forth in a recent monograph, [6] the concept of culture, the sum total of man's social heritage, had indeed been variously adumbrated but not clearly envisaged before Klemm and not rigorously circumscribed before Tylor's classic definition in 1871. Then at last it became obvious that culture embraced everything transmitted not by biological heredity, but by membership in a social group--not merely houses, crafts, and food-gaining pursuits, but also social structure and usage, religious ceremonial and belief, folk tales and amusements. This totality could no longer be adequately studied by the way; it required a specialist's concentration.

As a matter of fact, if a geographer or biologist could not do justice to the whole range of these phenomena, neither could an ethnographer. It developed that the aims of science could be best served by further specialization and that certain phases of culture could be adequately investigated only through collaboration. On the Cambridge expedition to Torres Straits organized by A. C. Haddon in 1898, each participant devoted himself to aspects of native life for whose study he was specially prepared: Haddon, a zoologist, took up material culture; Rivers, a psychologist with special interest in social life, gave psychological tests and applied himself to the intricacies of kinship; and so forth. Generally speaking, a full ecological picture implies the cooperation of a geographer, a botanist, and a zoologist; the songs, a feature often of great moment to the natives, must be transcribed and comparatively dealt with by a musicologist; the very nature of whatever metal work may be observed can be determined only by a metallurgist. To take the last instance, an ethnographer wants to know how the metal work of aboriginal Colombia compares with that of Peru. Only a trained expert can clarify the cultural issues: Did the Peruvians achieve bronze by deliberately alloying copper with tin? If so, what were the relative proportions? Did the Colombian "tumbaga" rival the Peruvian bronze in hardness and efficiency? For the answer the ethnographer obviously must depend on the chemist and metallurgist. [7]

Indeed, the ethnographer must at times rely on the aid of laymen. Since he is rarely able to spend more than a limited

time with the subjects of his study, various phases of native
life elude him. An intelligent missionary, trader, or squaw
man who speaks the native language may be superior in some
respects to the greatest scientist. In December, 1832, Charles
Darwin encountered the Tierra del Fuegians, leaving us a
good report so far as it goes. But beyond externals he could
not go. Says he: ". . . it was singularly difficult to obtain
much information from them concerning the habits of their
countrymen . . . it was generally impossible to find out, by
cross-questioning, whether one had rightly understood anything which they had asserted. "[8]

On the other hand, the most intelligent outsider or even an
ethnographer who is not abreast of recent developments in his
science is bound to fall short of the goal. To cite an egregious
illustration, excellent lay writers have described "uncles"
and "cousins" as fulfilling certain social duties, whereas any
novice today would determine the exact nature of the relationships involved. Some scientific problems defy solution precisely because observers who once had the opportunity to record certain facts neglected to do so. An approximation to an
ideal solution lies in the close collaboration of a trained investigator with an alert permanent resident, the former di-
recting the latter's attention to points otherwise missed, the
latter supplying what can come only from a prolonged so-
journ.

Because the ethnographer deals with the total range of human activity as socially determined, he at times puzzles the
votaries of other scientific disciplines. Sometimes they have
raised the issue whether students of, say, folk tales ought to
rank as "scientists" at all; I remember a relevant challenge
at a meeting of the National Academy. The answer is obvious.
Science must take cognizance of the whole of reality. The individual scientist does not control the segment of reality he
has chosen for his subject matter; his business is to investi-
gate it in all its manifestations. The paleontologist who finds
a fossil human skeleton does not ignore the stone artifacts
beside it; and the ethnographer, whatever his personal pref-
erences, cannot on principle ignore aspects of man's social
existence that happen to have hitherto been dealt with by humanists or even by amateurs. The point is that when *he* deals

with them he approaches them as an objective investigator.
Of course, he should be able to share the native's thrill over
an effectively told narrative; but as a scientist he must do
more. He must analyze it, compare it with similar tales elsewhere, determining its distribution, must establish its func-
tion--whether it is told for sheer esthetic entertainment or
as a sacred myth, whether it appears in several versions,
and so forth. If humanists have devised suitable techniques
for such purposes, the ethnographer will naturally make them
his own.

Put more broadly, the ethnographer in principle strives to
learn everything he can about all phases of culture; and in
order to do this he is naturally obliged either to devise appropriate techniques for study or to utilize the services of
experts from other branches of learning. The resulting attention to minutiae inevitably makes much of ethnography un-
palatable to the general public. Indeed, it becomes tedious
for the ethnographer himself so far as he is not specifically
interested in certain subjects. A description of textile weaves
or a roster of kinship terms is deadly unless it is vivified by
a sense of problem. The justification for delving into these
details lies in the fact that they can shed light on such broader
issues as human inventiveness and questions of genetic affinity.

To illustrate concretely what was involved in the coming of
age of ethnography as a science, we may consider the treatment of certain crafts by Georg Schweinfurth in the early
1870's. Schweinfurth, a justly famous naturalist and explorer,
was particularly interested in technology, and, on his most
celebrated expedition, he paid special attention to the Mangbettu (Monbuttoo) in what is now the northeastern Belgian Con-
go. Yet all he notes about the pottery of this people is that
it is handmade, superior to that of other African Negroes,
uniformly spherical, and, except in a unique specimen, lacks
handles, yet is ornamented with raised patterns that facilitate
manipulation. By way of contrast, in the beginning of the present century Mr. and Mrs. Routledge, amateurs themselves
though in close contact with professionals, visited an East
African tribe and in their report devoted five printed pages
and ten plates to the manufacture of pottery. They tell us that

women produce all the earthenware, that they get their clay
and other ingredients in such and such localities; we also learn
precisely how they moisten, mix, mold, and fire the clay in
order to achieve a usable utensil. [9]

Again, .Schweinfurth saw Mangbettu burden baskets, which
reminded him of the Thuringian_*Kiepen,* slung as they were
from the back by forehead or shoulder bands. He does not tell
us how the baskets are plaited, whereas a generation or so
later any tyro, though uninterested in technology, would bandy
such technical verbiage as "twilled openwork twine" or "three-
rod foundation. " The tyro would not approach Schweinfurth
as an observer, but his eyes would have been opened to the
potentialities of basketry for ethnographic research. An American technologist, Otis T. Mason, had in the interim produced
a standard work[10] classifying the weaves found in the collections of the United States National Museum, had determined
the distribution of techniques in the New World, and had referred to striking parallels in other parts of the globe. The
conditions favoring certain processes had been clarified and,
in most such studies, the alternatives of genetic relations
between distant groups or independent evolution had been presented or implied. What does it signify if "wrapped" basketry
turns up in southern Arizona, in archeological sites of the
Mississippi Valley, and in the Andaman Islands? Why are
Tierra del Fuegian baskets coiled in a manner strikingly suggestive of Asiatic types? Evidently it became obligatory to
consider the minutiae of hand-weaving. But corresponding
conclusions were thrust upon ethnographers whenever any
phase of culture was closely examined. Even a game of chance
might bear crucially on the connection between Asiatic and
American peoples. In short, on principle the science of ethnography could not neglect any trait, no matter how trivial
apparently, that human groups handed down from generation
to generation.

Another conclusion had to be rediscovered. The early theo-
rists had not confined their studies to primitive tribes, but
drew upon sources from ancient Egypt and China, classical
antiquity, medieval and postmedieval Europe. They realized
that culture formed an indivisible whole; its significant features could be ascertained just as well in the most complex

as in the rudest societies, in some respects better since the complex societies offered written documents and datable events. It was, from all points of view, artificial to segregate the nonliterate from the literate peoples. Ethnography had to turn into a study not only of all primitive cultures, but of all cultures past and present; it had to integrate findings among "savages" with those of prehistory and of the cultural researches among all civilizations. Even though the individual might control only a tiny corner of this vast field, he had to be cognizant of its essential unity and continuity.

As ethnography matured, it further became obvious that as a science it had to attain objectivity, the same quality that distinguishes the "historical-minded" historian from the partisan chronicler. Such objectivity was easy enough so long as the ethnographer investigated tools and implements, but became difficult in contemplating exotic practices, beliefs, artistic productions--in short, *values*. The temptation is very great to view these aspects of primitive or alien cultures ethnocentrically. Even far into the nineteenth century, scholars were unable to rid themselves of the notion that the norms of Victorian Europe were absolutely supreme and that any culture had to be graded in proportion as it approached the ideals of contemporary western Europe. So able a writer as Lord Avebury (Sir John Lubbock) was forever expressing indignation over the "revolting" practices of savages, whom he found "almost entirely wanting in moral feeling, " a conclusion he left unaltered in the 1911 edition of a work first published in 1870. [11]

Two comments seem apposite. In the first place, even by our norms, behavior that strikes us as outrageous reveals on closer scrutiny wholly unexpected psychological correlates. It is shocking that the Eskimo will abandon an aged parent. However, the act does not involve the callous brutality it suggests at first blush. The elders simply cannot keep up the pace required for the survival of the household and themselves beg, against the remonstrances of their children, to be left behind, lest everyone perish. Such instances have made us wary about ethical verdicts on the strange ways of "savages" or foreigners. But, second, the ethnographer, *qua* ethnographer, has no absolute norms for judging good and evil. As in the

Eskimo instance, he will always correct misconceptions concerning facts, and the truth will revise the uninformed judgment of fair and intelligent laymen. But he has no professional
competence to establish ethical canons. As a human being, he
merely hopes that his professional knowledge will create a
greater tolerance of spirit.

Because as a specialist the ethnographer accepts all cultures, some members of the guild have been sorely troubled.
Must they, then, accept Nazi Germany or Soviet Russia? The
difficulty is a purely semantic one. Ethnography "accepts"
Hitlerism or Stalinism as the zoologist "accepts" the skunk,
the rattlesnake, the mosquito, the turtle dove, and the ele-
phant. All of them are parts of the universe, hence science
must study them; their relative chastity, benevolence, and
malevolence simply do not enter.

This elimination of value judgments does not do away with
the study of values, of course. That certain societies prize,
above all, martial valor while others exalt the arts of peace
is an important fact, the relevant attitudes constituting diag-
nostic features of the associated cultures. As Kroeber puts
it, the science of culture deals with "values as natural phenomena occurring in nature."[12]

A philosopher unusually conversant with ethnographic literature has suggested that a science of culture which deprecates
the creation of ethical norms thereby confines its efforts to
the descriptive level. Nothing could be farther from the truth.
As has been hinted above, out of the raw factual material there
arise spontaneously efforts at interpretation. As in biology,
the mere phenomena of geographical distribution evoke an indefinite number of questions. In part they precisely parallel
those familiar in other disciplines. What is the significance
of cultural parallels? Are they the result of genetic affinity?
Are they due, in popular parlance, to similar causes operating
in similar circumstances? Have, perchance, diverse causes
led to convergent developments? How can we distinguish between homologous and analogous resemblances? Instead of
the cruder hit-and-miss correlations often suggested on intuitive grounds, E. B. Tylor long ago proposed a painstaking
recourse to statistics; and his pioneer effort has found follow-
ers in the writings of G. P. Murdock and Harold E. Driver--

irrespective of differences of interpretation as to specific points.

To summarize, the development of scientific ethnography conforms in broad outline to the formulation by Ernst Mach of all scientific development: The initial adaptation of thinking to the facts has been followed by the adaptation of thoughts to one another. Or, put in other words, observation has been followed by integration.

At this point it may be well to consider the kind of integration so far achieved and likely to be attained in the future. If it holds true generally that no branch of learning is under obligation to ape the procedures of any other, this may be asserted most emphatically with reference to the search for ultimate explanations. For that reason it strikes me as a momentary relapse into unwarranted subservience to an older discipline when Professors Kroeber and Kluckhohn in their recent treatise on "Culture" wistfully contrast the theory of gravitation with the lack of a "full theory of culture" and declare that "we have plenty of definitions but too little theory. "[13] Certainly we have too little sound theory; and because cultural phenomena are incomparably more complex than those of physics, because they have come to be studied at a much later period, it seems unreasonable to expect the same measure of integration either now or in the future. As Thurnwald aptly contends, the web of sociocultural phenomena is too intricate to permit a resolution into all its components without residue, yet we must attempt to discover all possible connections and determinants. [14]

What is the nature of such interpretations? In ethnography, integration has been historical or processual, though as a rule neither type has excluded the other. When investigators have stressed the historical point of view, more or less as biologists did in the era of phylogenetic speculation, they have not neglected the factors that make for cultural change. This is equally true of those generalizers who believed in uniform stages of evolution the world over and of the more recent scholars who postulate development along several distinct lines. For example, in 1877 Lewis H. Morgan outlined a series of statuses from the lowest savagery to civilization and regarded progress as "substantially the same in kind in tribes and na-

tions inhabiting different and even disconnected continents";
but he was by no means content with tracing the sequence of
events. Thus, in strongly maintaining that patrilineal descent
was everywhere preceded by matriliny, he offered a cause for
the shift, namely, the rise of property along with the desire
to transmit it to one's own children rather than to matrilineal
kinsfolk. [15] Father Wilhelm Schmidt, one of Morgan's most
determined critics, opposes to Morgan's unilinear system a
multilinear scheme of development and rejects the theory of
matriliny as a necessary antecedent of patriliny, assigning
both rules of descent to originally distinct specializations of
the primeval state of sexual equality. But in order to account
for the rise of either he has recourse to the economic pre-
ponderance of the female and the male sex, respectively. [16]

The double interest in a chronological arrangement of the
facts and in an explanation of why they appeared conjointly or
in the succession discovered characterizes some of the most
recent theories. V. Gordon Childe sees in the invention of the
wheel one consequence of metal tools in the hands of skilled
craftsmen; and he shows how the substitution of iron for bronze
democratized metal work, inaugurating a series of technological improvements that culminated two thousand years ago in
"all major manual tools of industry and agriculture" known in
the Mediterranean area. [17] J. H. Steward consciously joins
sequential and functional considerations. He traces the development of higher civilizations in the Old and New World to
the invention of irrigation. This produced in both cases an increase of population; and "a priestly class developed because
increasing populations, large irrigation works, and greater
need for social coordination called upon religion to supply the
integrating factor." [18]

Given the diversity of cultural phenomena, it is hardly to be
expected that all of them should be at any one time amenable
to the same degree of analysis. Apart from language, which
represents a highly specialized type of socially transmitted
features, social organization probably represents the department of culture within which functional relationships have been
most satisfactorily defined. Such relationships have often been
alleged, probably from the beginnings of ethnographic observation in its prescientific stage, but with little attempt to sub-

mit the proposed correlations to a rigorous test. A famous lecture by E. B. Tylor dating back to 1888 marks an epoch, irrespective of the acceptability of some of his conclusions, for it represents an effort at exact statistical verification. [19] For example, Tylor asks whether the frequent association of a taboo against speaking with one's mother-in-law and the practice of settling with the wife's parents is due to mere chance or to an organic nexus of the two customs and arrives at a positive conclusion, though explicitly recognizing the existence of more than one determining factor. The most ambitious attempt to follow in Tylor's footsteps is a book by Murdock, [20] in which numerous correlations are defended on a statistical basis. Though a recent critic considers Murdock's correlations too high, he admits that "their value taken collectively is beyond question" and wholeheartedly concurs in the use of exact treatment of our observational material. [21]

It should not be supposed that sound interpretation in this field hinges solely on specific mathematical techniques. Even without them Radcliffe-Brown, Lévi-Strauss, Eggan[22]--to mention some conspicuous examples--have shown that certain phenomena of social structure are intimately related, which of course, as Tylor already realized, does not mean a simple 100 per cent causal nexus. The results are due to a rigorous definition of concepts and a broad comparative approach. None of these writers would claim to have found a full theory of social structure, but as Eggan remarks: "Generalizations do not have to be universal in order to be useful."[23] In the processual sphere the progress of ethnography lies in the ever-increasing and increasingly better-founded determination of functional relationships between descriptively isolable elements, in other words, in constantly detecting more and more organic bonds between apparently disparate phenomena. A single formula for the cultural universe, a simple solution of its enigmas, would prove a snare and a delusion.

On the historical side, ethnography has assiduously determined an infinitude of special intertribal relationships and sequences. In broad outline, at least, the main steps have been traced by which the more complex civilizations of the Old and the New Worlds, respectively, have been built out of simpler beginnings. The whole story will never be known, but every synthesis of field discoveries with archeological research marks another step in the scientific progress of the discipline.

MAIN TOPICS
IN MARCO POLO'S
DESCRIPTION OF THE WORLD

BY LEONARDO OLSCHKI

If one would ask today "what kind of a man was Marco Polo?" the answer from experts, biographers, and reference
books would always be the same. It would state that he was a
Venetian merchant who traveled in Asia for twenty-five years,
mostly in the service of the Chinese-Mongolian imperial administration. Taken prisoner by the Genoese shortly after his
homecoming, Marco Polo dictated his *Description of the World*
during his detention in Genoa in 1298 and died in Venice in
1324.[1]

In this simple statement, as well as in more elaborate accounts, the only personal qualification of the man is his being
a merchant. In reality this is one of the most misleading errors commonly made in the appreciation of Marco Polo's book
and personality. As a matter of fact, he was never engaged
in any trade whatsoever. While in Asia he first traveled in the
retinue of a papal embassy entrusted to his father Niccolò
and his uncle Maffeo, and, later on, as an imperial functionary
and envoy to China's most remote provinces and borderlands.

Back in Venice after his long absence, Marco Polo did not
settle down as a businessman, or ever make a conspicuous
fortune as a result of his activities, travels, and commercial
experiences. He seems to have participated with other Venetian patricians in a naval expedition against the Genoese as a
gentleman-commander of a galley equipped with family funds.
In his home town he made investments of different kinds, as is
still quite usual for men not professionally engaged in spe-

cialized commercial activity. None of the documents at hand
mentions him as a merchant *(mercator)*, but always and only
as a nobleman *(nobilis vir)*, without any specification of trade
or regular activity in public life.

He became a popular figure as a storyteller, although some
people did not trust his veracity. Nevertheless, Friar Francesco Pipino translated his book into Latin at the suggestion
of the General Council of the Dominican Order and recom-
mended its perusal by the faithful who "may merit an abound-
ing grace from the Lord."[2] And indeed, Marco Polo felt and
presented himself therein as a lay missionary and champion
of the faith who set out for China with papal credentials and
some oil from the lamp burning on the Holy Sepulcher, in
order to prepare the conversion of Qubilai Kaàn, the emperor,
and the triumph of the Catholic Church among his vassals and
subjects in the Far East.

But his book was not intended to promote missionary or commercial ventures, or to be read as an edifying treatise showing
in its wonders and marvels the power and wisdom of God. Marco Polo composed his work as a "Description of the World"
for "great princes, emperors, and kings, dukes and mar-
quises, counts, knights, and burgesses, and people of all
degrees who desire to get knowledge of the various races of
mankind and of the diversities of the sundry regions of the
world"; in other words, a popular book for universal lay readers already acquainted with a vernacular geographical literature showing the wonders of the East.

The substantial difference between the contemporary common geographical lore and Marco Polo's *Description of the
World* consists in the empirical character of the latter and
the bookish origin of the others. With their praise of the Lord
and professions of faith, all of these books compiled in the
thirteenth century were intended for lay people who wanted
to share the knowledge of the world with the scholarly, Latin-
speaking teachers of letters and science.[3]

Just as in this medieval didactic literature, commercial
data are not very frequent in Marco Polo's book. Much in
contrast to the famous *Pratica della mercatura* of his younger
Florentine contemporary, Francesco Balducci Pegolotti,[4]
commodities and all kinds of merchandise appear in the Vene-

tian's report as characteristic aspects of the life of Asiatic
countries and peoples, rather than as objects of interest for
trade and export. Marco Polo had almost no experience in
this field when he left for China with his father and uncle at
the age of seventeen. Even the elder Polos stopped working as
merchants from the very moment they became ambassadors
of the Great Kaàn in 1265 and later on dignitaries of his cos-
mopolitan court. Commercial activity was incompatible with
their rank in the emperor's retinue and administration.

Consequently young Marco learned from them and from the
peculiar circumstances of his amazing career to take a much
broader view of the world than any medieval merchant on the
roads and in the markets of Asia. Nevertheless, this view was
a personal one, limited by a rudimentary culture and the contingencies of his official employment. It is also blurred by the
lack of scientific rigor and methodical consistency, qualities
which were scarcely fostered or appreciated before the seventeenth century in any branch of science and learning.

There is in his book a striking contrast between the circum-
stantial diffuseness of edifying and entertaining tales on the
one hand, and the parsimony and inaccuracy of most of its
geographical details. Although the Venetian traveler honestly
strove to compose an objective report of things seen and heard
in twenty-five years of incessant voyaging by land and by sea,
he omitted many characteristic aspects of the continent and its
peoples, as for instance, the extensive use of tea in the Far
East, the Great Wall of China, book printing, firearms, and
the compressed feet of the Chinese women he so much admired
as "angelical creatures."

These shortcomings may be explained and excused by Marco's deathbed confession to his friends that he had not told one-
half of what he had really seen. One could add many more
aspects of Asiatic life and nature unknown in Western countries
on which he was unwilling to elaborate. Still, there are enough
in his report. But owing to the very character of his book,
which is an itinerary and at the same time a comprehensive
treatise of empirical geography, the countless data on the
most varied facts and subjects are scattered within the personal narrative of recollections, impressions, and adventures.

Thus a substantial and critical re-evocation of the environ-

ment and experiences of the great traveler is possible and
satisfactory only if the facts and allusions contained in his
book are collected and coordinated in a system of topics.
Strangely enough, this has never been attempted although
Marco Polo himself pointed out that his book dealt with "the
people, the beasts and birds, the gold, silver, stones and
pearls, all sorts of merchandise and many other things" he
had seen and heard of when traveling in the East. [5]

This is, indeed, a very sober and quite inadequate enumeration of the topics he covered in the book. But as main objects
of the author's intellectual and practical curiosity these few
concepts may serve as a starting point for a more articulate
classification of matters considered in what he called a *Description of the World*. It is at once evident that Marco Polo
was interested mainly in the human aspects of Asiatic life,
and, in addition, in animals and some products of the soil
with intrinsic and commercial value, such as precious metals
and stones. The rest is simply called merchandise or vaguely
hinted at as "many other things."

In so doing, the great traveler really neglected the very
topic for which he is most praised: that is, the geography of
Asia, the scientific value of which has been disclaimed by
professional scholars for centuries and still is problematic
because of the vagueness of the data and the errors in report-
ing distances and in locating the places mentioned in his book.
His description of the Eastern countries is indeed related to
his personal and official affairs and to travel experiences
rather than to the results of systematic inquiry and scientific
investigation. Consequently most of his geography is personal
reaction and recollection of itineraries, although there are in
his book a few allusions to the maps used by mariners in the
Indian Ocean, [6] but hardly by merchants on the ill-marked
caravan routes of continental Asia. On the other hand, while
data on roads and means of communication are scanty or totally absent, he gives a minute description of Persian and Chinese
ships which he was able to judge more or less favorably as a
Venetian expert in this field.

The appraisal of these rhapsodic contributions to geographical lore started little more than a century ago, when historical
interest, a revaluation of empirical knowledge, and the sys-

tematic exploration of central Asia led to a more correct estimate of the book and of the author's unprecedented achievements. Since the fourteenth century some map makers had
profited from his account in tracing a more detailed, if still
fantastic, picture of the continent. But only one of the leading
scholars of his era availed himself of Marco Polo's experiences for scientific purposes. The physician and astrologer
Peter of Abano, a celebrated professor in Paris and Padua
early in the fourteenth century and one of the most famous
victims of the medieval Inquisition, consulted him in order
to get reliable information about the antarctic polar sky and
direct evidence of the habitability of the southern hemisphere,
then doubtful or denied by representatives of medieval science. [7]

As a result of the information he received, Peter of Abano
was able to write in his *Conciliator,* published in 1303, that
"Marco the Venetian, the most extensive traveler and the
most diligent inquirer whom I have ever known, " saw to the
south of the tropical zone, inhabited by wild men, "a star as
big as a sack, " which "had a faint light like a piece of a cloud"
and "the Antarctic Pole at an altitude above the earth apparently equal to the length of a soldier's lance, whilst the Arctic
Pole was as much below the horizon. "[8] This inaccurate statement about the so-called Magellan cloud and the antarctic
constellations, well known to Arabic astronomers, globe makers and mariners of that time, was their first direct mention
contained in a Western text.

But Marco seldom looked at the stars and never seems to
have traveled by means of quadrants and astrolabes. This
might have prevented him from anticipating the astronomical
blunders made almost at the same time at the southern tip
of India by Friar John of Montecorvino, first bishop of Peking, [9] and two centuries later by Columbus in equatorial regions, or those attributed to Vespucci in describing the Antarctic sky. [10] The Venetian preferred to study human geography
when aspects of life and nature roused his interest or curiosity.

He therefore mentioned the customs of peoples as the first
topic of his book. By this comprehensive term he evidently
meant what we call a national or tribal civilization in the
broader sense of the word. It includes religion with its cults,

rites, institutions, and magic practices, but without any consideration of its spiritual background or of its most elementary
principles of faith and doctrine. This category also includes
the traveler's data on the political status of every people and
tribe, their forms of government, their languages, trades,
customs, history, handicrafts, and products, with some mention of the peculiarities of their character, traditions, and
folklore, while cultural endeavors and achievements are systematically ignored.

These records are sometimes dry and sketchy, as in the
mention of places in Persia, Turkestan, and most of southern
China; at other times they are elaborate descriptions enlivened
by tales and anecdotes, as those contained in the chapters
devoted to western Asia, Cathay, Tibet, Japan, and India,
and in reports about the principal towns of China, and on the
marvels of insular and tropical Asia and the Indian Ocean
as far as Socotra and Madagascar. The religious life of all
these countries is his main concern. He hates and despises
all their aberrations and devilries, but he seems to have overcome his genuine Catholic feelings and superstitious fears
in intercourse with the natives, and in his fairly detached
mention of their beliefs and rites. As a man of the world and
a lay missionary of the faith, Marco Polo never exhibits the
stubborn and defiant attitude of those Dominican monks who,
in 1247, imperiled their mission in Tartary, [11] or of the Fran-
ciscans who provoked the Muslim of Morocco, in 1227, [12] and
of those who, in India, in 1321, paid with their lives for their
irate contumelies against the Prophet and his law. [13]

Thus the Venetian traveler was able to draw up a reliable
and almost complete denominational geography of Asia with
the same detachment that enabled him to sketch a picture of the
political situation of the continent in the last decades of the
thirteenth century. In so doing he distinguished the countries
subject or tributary to the Great Kaàn from the independent
territories still on the agenda of the Tartar conquest of the
world. As a loyal functionary in the imperial administration,
as an objective observer of tribes and nations still restive
under the Tartar sway, and finally as a faithful member of
the Sino-Mongolian nobility, he assumed a legitimistic attitude
even in cases when the Christian cause had to yield to the

rights of a pagan sovereign, and even when his stand contributed to the conquest and subjection of peoples yearning for
political freedom and cultural independence. This same attitude, by the way, would have been displayed by any Venetian
nobleman toward the policy of conquest and subjection followed
by the Most Serene Republic of Venice in Marco Polo's day.

Our traveler was, therefore, a keen observer of the highly
disciplined military organization of the empire on which the
power of the Genghis Khan dynasty rested in the first century
of its development. His description of the military hierarchy
and strategy, of arms and cadres, of warfare and battles had
a forerunner in the report of Friar Giovanni da Pian del Càr-
pine, first papal envoy to the Great Kaàns, in 1247. [14] But
while this clever and brave missionary intended to teach the
Western leaders how to organize the defense of Christianity
against new Tartar invasions, Marco Polo described their
military system as the main aspect of their civilization and
as a proud and powerful institution to which he temporarily
belonged as a Sino-Mongolian knight.

In his legitimistic and bureaucratic attitude toward the subjects of the Great Kaàn, Marco Polo shared the conquerors'
indifference and suspicion with regard to the cultural achievements of such highly civilized Asiatic countries as Persia and
China, which were then parts of the Mongolian Empire. He
was, of course, well aware of the wealth and splendor of these
exotic civilizations and did not fail to praise their noble cities,
monuments, palaces, walls, towers, bridges, and castles,
describing their shape, dimensions, material, and ornaments.
But in doing so our traveler is more concerned in exalting the
Kaàn's glory, power, and wealth, than in appreciating the skill
and refinement of native architects, painters, and craftsmen.

Yet Marco Polo's book contains many scattered data on these
oriental monuments worthy of collection and critical explanation. They reveal his natural interest in architecture and decoration, characteristic of the artistic background of his Venetian culture and experience. Conversely, his consistent reticence with regard to the highly developed poetry and music
of all the countries described in his book shows his preference
for the more striking and ponderous expressions of artistic
life, even if less characteristic, genuine, and refined.

Similarly, while he mentioned churches, mosques, temples, sanctuaries, monasteries, priests, monks, ceremonies, and liturgical implements of all kinds, he ignored the tenets on which the Nestorian, Mohammedan, Buddhist, Confucian, Taoist, and Hindu civilizations were based, even when the emperor took advantage of all these creeds and philosophies to increase his power and to strengthen his influence in every part of his huge, motley domain. [15]

Following this trend, Marco Polo took a lively interest in what he and his contemporaries considered the leading sciences and liberal arts, namely, necromancy, astrology, geomancy, and physiognomy, all regarded as complementary branches of the practice of medicine. For all these disciplines studied and practiced in Asia he mentioned two principal centers: Baghdad, the capital of Irak and once the seat of the Caliphate suppressed by the Mongols, and the city of Soochow in southern China, whose famous school of medicine and philosophy enjoyed a high reputation throughout the country. All the great centers of Asiatic culture situated between these two border towns of the empire had been destroyed by the Tartars and never recovered from the blow. [16]

Our traveler was well aware of the difference between these fields of learning and professional activity on the one hand, and, on the other hand, the magical practice of the Mongolian shamans, the Tibetan bakshis, the wizards of Kashmir, the Taoist sages and alchemists, and the exorcists of India--all extensively active at the Kaàn's court and in the public and private life of their native countries. Despite his Christian horror of these "devilries," he was somehow fascinated by such magic performances, to the point that he once availed himself of these pagan practices when he lost a precious ring. Indeed, he recovered it with the help of some idols which he afterwards tricked by denying them the due offerings and homage. In doing so the shrewd Venetian transformed a mortal sin into a piece of fun and a fact-finding experiment. [17]

He would not, however, have dared to make fun of these four branches of medieval science which he recognized as characteristic expressions of oriental civilization. All of them had reached in his time the dignity of leading disciplines, teaching the various modes of divination and becoming in the

East as much as in the West instruments of political power, of scientific knowledge, of healing practice, and of occult virtues. Just as every contemporary Western ruler appointed skilled astrologers and soothsayers to direct his policy by predictions, Qubilai Kaàn allegedly employed five thousand Christians, Saracens, and Cathayans, "who investigated the course and events" of the whole year according to the practice of each sect. While in China the annual compilation of pre-monitory calendars was the emperor's chief responsibility and the most prominent state action, the art of divination had attained in Italy, under governmental and courtly protection, the most extensive theoretical development and the greatest popular favor.

This was the only field of knowledge and activity in which the West and the East merged in those days, including in this world-wide agreement all creeds, peoples, sects, and schools. All these branches of scientific divination were merely adjustments and ramifications of doctrines already elaborated in eastern Persia in the era of Alkindi, Alfarâbî, Albumazar, Avicenna, and other more or less famous masters in the theory and practice of occult science and the power of the stars. They had expanded from there westward toward Egypt, Sicily, and Spain, and eastward to Mongolia and China where they finally converged with Indian, Tibetan, and indigenous animistic and shamanistic trends and traditions of esoteric science and religion.

For Marco Polo and his contemporaries this was science at its highest fulfillment. He was unaware of his own scientific inquisitiveness when he observed natural phenomena on the earth or in the sky; for instance, when he measured in his own way the height of the polar star over the equatorial horizon, or when crossing the Pamir he noticed that at those altitudes it took longer to cook his meals than down in the plateaus and plains of Asia. He attributed the latter phenomenon to the cold, unaware like everyone else, of course, of the physical problems involved in this experience.

Like most of his contemporary scientists and laymen, Marco Polo was less interested in the ordinary ways of life and nature than in their exceptional, strange, and curious appearances. Therefore, while mentioning a hundred times the countries and places where people lived by handicraft and trade, he usually failed to note the details of these activities, but gladly reported the striking peculiarities of customs in fashions and ornaments, of funeral rites and matrimonial usages, of superstitions and conventions, of the most salient aspects of local folklore, such as tattooing, ritual cannibalism, the *couvade,* and the oddities in food, drink, sexual intercourse, and domestic life.

Certainly the aspects of nature and the products of the Asiatic soil make up the bulk of the book. Indeed, a simple statistical account of the animals mentioned in his *Description of the World* would confirm Marco's keen interest in the animal life of Asia. It was extended not only to mammals and all kinds of birds, but included fish and reptiles as well. This is a peculiar feature of his book, which reflects in this field of interest and experience the general curiosity about animal life characteristic of the thirteenth century, as manifested in Western art and decoration, in courtly literature and vernacular texts, and in scientific endeavors stimulated by the translation of Aristotle's *Historia animalium,* and even in religion, morals, and philosophy. While this general trend led the Venetian traveler to observe "the beasts and birds" of Asia, his curiosity was enhanced by the sight or the tales of wild, exotic, and fabulous animals like lions, tigers, elephants, camels, bears, unicorns, griffins, and whales, always considered wonders of the East.

With his eyes sharpened by a revived interest and long experience, Marco Polo discovered animal varieties unknown to his Western contemporaries, as for instance the famous *Ovis poli* of the Pamir, the "wonderful" yaks of northwestern China, the wild beasts of Tibet, the musk deer of Kansu, the sacred cranes of China, the most varied sorts of game everywhere in the continent, besides the fowls of Fukien and the rich fauna of the tropical regions of Indonesia, Burma, and India. He was so much impressed by all these zoological discoveries that even the most familiar types of animals might elicit an expression of enthusiastic admiration, as for instance the horses of Turkmenistan and Badakhshān, the wild asses of Persia, the cattle of High Asia, the falcons of Mongolia, the pheasants of China, the game in the Great Kaàn's hunting grounds--in short, whenever and wherever he saw them with the eyes of a passionate sportsman and an experienced connoisseur.

It can be taken for granted that Marco Polo did not develop these qualities in his native Venice, where he could have seen at best some horses, donkeys, dogs, cats, and rats as the main living representatives of the animal kingdom, besides those appearing in the famous ancient and medieval sculptures all around St. Mark's basilica. He became familiar with the Asiatic fauna mainly in the Mongolian environment, where a people of shepherds and hunters had always lived in symbiotic community with all kinds of animals.

This typical feature of a nomadic society of steppes and forests is impressively represented in the first and main book of their tribal traditions, the famous *Secret History of the Mongols* composed in 1240. It was adhered to for generations in the courtly practice of the Mongol dynasty throughout the continent, and by the Mongol nobility which had become sedentary in settling down in the highly civilized countries of eastern and western Asia. Marco Polo was connected with both the courtly and ruling society and acquired in this environment his familiarity with animal life. He had his main training in this field of his Asiatic experience through direct participation in the frequent and huge imperial hunting parties minutely described in several chapters of his book. He became an expert in the noble art of falconry, which had developed in his day into a main topic of courtly and learned literature after the Emperor Frederick II, his son Manfred, the century's greatest scholar Albertus Magnus, and Dante's master and friend Brunetto Latini had given this branch of activity the dignity of a science. [18]

These strategically organized hunting parties and the vast animal parks around the Kaàn's residences became the school in which Marco Polo learned to understand and to describe the "beasts and birds" mentioned in his book. By collecting these records, a substantial encyclopedia of Asiatic zoology could be compiled. It would turn out very much unlike the contemporary literature of this kind, which was mainly based on learned traditions and legendary fancies, and was seldom enlivened

by that touch of life and realism that makes Marco Polo's
dry report attractive even to a modern reader. [19]

The same relationship between learned traditions and professional concerns on the one hand, and direct observation
and personal curiosity on the other, can be found in Marco
Polo's attitude toward the varieties of Asiatic vegetation. Even
in this field there are in his book several allusions to the bookish literature which had become in his day popular and ubiquitous through vernacular compilations and textbooks for
schools. To start with a striking example, one of the most
popular landmarks in the legendary visions of Asia was the
so-called "Dry Tree" or *Arbre sec,* also called the "Solitary
Tree" or even the "Tree of the Sun and the Moon." In Western
tradition its story goes back to the poetic history of Alexander
the Great, concocted in the fourth century A.D. by an unidenti-
fied Greek author known as Pseudo-Callisthenes and widely
known in Christian Europe as well as in the Islamic world
from Spain to Indonesia. Marco Polo first became acquainted
with this christianized pagan legend from Italian storytellers
in his home town, and, later on, through local information
collected when traveling in Mohammedan countries. Finally
he found the legendary tree near the northeastern Persian
border in the province of Tonocain and gave a description of
it so reliable in all essential details that botanists were able
to identify it as the Oriental plane, or chinar, which gives a
touch of life and green to the desolate landscape of that part
of Asia. [20]

In that case a literary reminiscence preceded a botanical
discovery, showing that the Venetian traveler could look at
trees and plants independently of officinal, professional, and
commercial interests in the flora of the continent. In fact, in
his long wanderings through most of its regions he noticed,
sometimes with surprise and admiration, the boxwoods of
Georgia, the palm groves of Persia, the pine trees of Mongolia, the big forests near the Kaàn's summer palace of Shang-
tu, the huge plantations of mulberry trees and the bamboos
of China, the endless rows of umbrageous trees along the
imperial highways, the exotic vegetation of Champa and Sumatra, the luxuriant artificial gardens of Alamut in the Elbrus

mountains, the imperial parks of Peking and Quinsai, the orchards of Khorasān, the vineyards of Kashgar in Turkestan, the canes of Tibet and many other plants of all kinds in every region of central, insular, and tropical Asia.[21]

In the same way Marco Polo was the first Western traveler able to localize the different sorts of spices which had always been, but especially in his era of commercial enterprise, one of the most valuable commodities in demand, and at the same time the fragrant and mysterious symbol of Asiatic wonders and wealth. Since the export of spices from the East to Mediterranean markets had been for centuries an Arab trade monopoly, nobody knew their exact origin. Marco finally revealed to his contemporaries that aloe was abundant in Indo-China and Indonesia; that cinnamon was produced in Tibet and Malabar, gelenger and ginger in southern China, cloves in the inaccessible mountains of the western Chinese region of Kienchang; that most of the pepper came from Bengal, Sumatra, and some other parts of tropical India, while rhubarb was mainly cultivated in the Chinese province of Tangut (Kansu), from where merchants, as he says, spread it to all parts of the world.

Western traders, scholars, and consumers did not know even the names of these regions. From Marco Polo's book they all learned what clove shrubs, the coconut palm, and other exotic officinal plants looked like. With that the Venetian traveler became the first botanical explorer in history. In this case it was the commercial value of these drugs and vegetables that prompted him to look at them with an inquisitive mind. In doing so he intended to satisfy the curiosity of his fellow countrymen rather than to promote an invasion of Western merchants into those distant regions and uncharted land and sea routes. But the information contained in his book prompted the Portuguese to break the Arab trade monopoly and the Turkish trade barriers by organizing two centuries later their expeditions to India and the fabulous lands of drugs and spices, almost at the very time that Christopher Columbus drew inspiration from Marco's data in planning a conquest of the lands of the Great Kaàn and the spice islands beyond the ocean.

From time immemorial science and fable had located in the

East the production of gold, silver, gems, and pearls. Many
centuries of commercial experience had confirmed those tales
and reports. Accordingly, Marco Polo mentioned these valuable things as a third item in his list of the subjects to which
his book was devoted. Yet, despite his lively interest in them,
it was only in his reports on insular and tropical Asia that
these riches became a regular topic of his book. He mentioned
only incidentally the silver mines of Armenia, Badakhshān,
and Zachar, all inaccessible to Western exploitation and trade.
His interest in gold starts with his fantastic description of
Japan, a country known to him only by hearsay and Tartar
propaganda. From then on gold is frequently mentioned among
the riches and wonders of the East.

Its late appearance in the book is explained by the fact that
the output of gold was unimportant in China proper and in many
other continental countries tributary to the Mongolian Empire.
Moreover, gold never played an important part in the Chinese
economy and monetary system, based on silver and copper,
which were frequently replaced by paper money as the only
legal tender. [22] Marco Polo gave a good description of China's
currency and praised the emperor as the best of alchemists
because he was able to transform into gold even the bark of
the mulberry trees so abundant in his domain.

Likewise, he seldom yielded to the literary and legendary
fascination of the pearls, gems, and precious stones, of which
he mentioned diamonds, rubies, turquoises, topazes, sap-
phires, amethysts, and sundry other varieties known from
ancient sources and in medieval lapidaries as characteristic
features in the current picture of Asia. [23] Of their secret virtues he knew only what was told about a particular stone in
Japan that was said to make its wearer invulnerable. And
although he once repeated the old story of eagles carrying
diamonds concealed in pieces of lean meat, the data in his
report about the production of pearls and all kinds of gems
and precious stones appear for the first time in literature free
from the fabulous trimmings cherished by the merchants and
customers who traded the products of Asia.

The commodities and other objects of commerce mentioned
in Marco Polo's book include the products of all kinds of crafts

and trades and constitute one of the most varied and valuable
contributions to our historical knowledge of the medieval East.
Most interesting among them are the products of the Asiatic
soil described by our author. It is well known that his book
contains the first mention of naphtha in Western literature. He
discovered it, so to speak, in the region of Baku, then a bor-
derland between Armenia and Persia, where it was used for
burning and medical or veterinarian purposes.

Marco made a clear distinction between this mineral oil and
the vegetable oils drawn from sesame and walnuts in the woody
valleys of Badakhshān and Kashmir. Moreover, we find in his
report the first and most extensive description of the production of asbestos from the mines of Chingintalas in the Altai region, one of the richest in all kinds of minerals in continental
Asia. He was able to inform his contemporaries by direct
experience and reliable information that this substance was
by no means, as was generally believed, the wool of a fire-
proof salamander, but an incombustible and textile mineral
of which Qubilai Kaàn had a towel sent to the pope as a wrap-
per for Christ's holy handkerchief. [24]

From Marco the Western world heard for the first time
about China's extensive production of mineral coal, burned
for fuel throughout the country and so plentiful as to enable
a vast number of people to have a hot bath "at least three times
a week, and in winter if possible every day, whilst every nobleman and man of wealth has a private bathtub for his own
use. " Wood, although plentiful, would not suffice for the purpose.

These remarks are indicative of Marco's approach to Eastern civilization. He did not mention how important the production of coal had been for centuries in China's economy and
mining industry. But as a man coming from a country where
a hot bath was a costly luxury and a privilege of the few, and
then living among the Mongols, a people more afraid of water
than of fire, the Venetian was struck by the hygienic exploitation of coal by the Chinese, whom he probably despised as an
effeminate and rotten race.

Metals and minerals had always been in China both imperial
regalia and monopolized products of the soil. This must have
been the main reason that our traveler and imperial functionary

paid more attention to iron and salt than to gold, silver, and
precious metals extracted from oriental mines. Iron and salt
were the principal sources of state revenue, and the substances
most praised by the Mongols, who were unable to handle metals
and probably to produce salt in considerable quantity for domestic and animal use. So great was the appreciation of iron
among the nomads of the steppes that medieval legends of
eponymic origin made an ironsmith out of young Genghis Khan,
because his early name of Temudjin indeed designated a man
skilled in the handling of iron. After the conquest of most of
the continent, the Mongol rulers entrusted the administration
of the iron mines to foreigners and employed prisoners of war
and slaves from foreign countries in the increasing exploitation of the iron pits of central Asia. The fifty-five known iron-
producing districts of southern and western China under Mongolian administration were exploited by means of forced labor
under state supervision. [25]

Marco Polo's keen interest in this branch of Asiatic production and trade was not directed by commercial purposes. The
export or import of iron and other metals from those faraway
regions of the earth would have been beyond the means of the
Venetian or Genoese merchant of those days. How little our
traveler was influenced by commercial concerns in sketch-
ing his description of the world can be proved by his reticence
with regard to the two most coveted mineral products in the
contemporary intercontinental trade: tin from India and the
Malayan Peninsula, and alum from Asia Minor. Although China
imported a considerable amount of tin, and although both the
Venetian and Genoese traders were very active in the alum
business, Marco Polo never mentioned these remarkable
sources of his home town's wealth. [26]

It is very probable that the interest he took in iron and steel
was determined by his activity as a functionary in the imperial
administration of revenues and monopolies. Iron had become
extremely important for the conquering Mongolians who, before
the invasion of China and Europe in the first half of the thirteenth century, used mainly leather for their military equipment and only afterwards shifted to metal for the production
of their implements of war. Marco certainly was one of the

foreigners employed by the Mongolian rulers in this particular branch of their powerful bureaucracy.

The same circumstance explains Marco Polo's persistent and punctilious mention of such a common and ubiquitous product as salt. But salt was also for centuries the most important source of state revenue in China as well as in Venice, and at the same time the most powerful political instrument of both those distant domains in keeping subjects of conquered or tributary regions under their sway. [27] The state salt monopoly had been unpopular in China since Huan Kuan composed in the first century B. C. his famous *Discourses on Salt and Iron (Yen T'ieh Lun),* one of the masterpieces of economic literature of the ancient world. [28] This branch of the imperial administration was the most conservative of the whole Chinese bureaucracy. It was taken over by the Yüan dynasty and provided with foreign supervisors. Marco Polo must have been one of these functionaries when he resided for three years in the town of Yang-Chau, a center of Chinese salt production and administration.

There he must have acquired an extended professional knowledge in this field, to such a point that his exact data on the salt revenue match everything related by contemporary Chinese sources on this subject. It was with an expert's eye that he gave a reliable report on nearly the whole salt production of the Mongolian empire. And since the administration of the salines ranked very high in the imperial bureaucratic hierarchy, the top mandarins of this leading fiscal branch were sometimes granted political powers. It is because of this privilege that Marco could say, with little exaggeration, that he had ruled for three years the old and thriving provincial capital of Yang-Chau, one of the towns of southern China most reluctant to accept the new masters from the North.

Marco's native republic of Venice had a similar fiscal and political organization, as dignified and powerful as the Chinese was from earliest days down to the end of the imperial era in 1912. [29] Our traveler might have attained in the Far East a governmental office corresponding to that of a Venetian *provveditore al sale*. This circumstance would explain his keen

interest in a branch of Oriental life which other travelers
would have overlooked in favor of more spectacular aspects
of the Great Kaàn's fabulous empire. [30]

All of the many other topics of his narrative reveal the same
trend and spirit. They mostly deal with the natural and industrial products of Asia of which Marco Polo consistently
describes the origin, use, and distribution, for instance: the
kind of vegetables grown and consumed in the different regions
of the continent, or the ordinary commodities produced or
manufactured throughout Asia, such as cotton, wool, furs,
leatherware, silk, and all kinds of textiles, carpets, rugs,
and needlework. Of some of these items he brought samples
home to Venice.

The interest he took in the human, natural, and historical
aspects of those distant countries was directed by the phases
of his personal career. The information he gave his fellow
countrymen about Asia was influenced by what he knew would
appeal to his readers and listeners in the fields of religion,
government, and business. The vernacular language in which
the book was compiled and published indicates that it was not
intended as a scientific treatise of scholarly interest, but
rather directed to the lay world and concerned with facts and
events within the reach of its understanding. Nor was Marco
Polo ever merely concerned to satisfy the curiosity of people
who had long thought of those distant countries in terms of
the current teratological tales and preposterous fancies. [31]

With all its limitations and shortcomings Marco Polo's *Description of the World* is the first book of empirical geography
of modern times. It was a long time before empirical and
scientific geography merged into one branch of learning for
a better knowledge of our mother earth. Since then the scholarly appreciation of Marco's book has grown in proportion to
the development of geography into a universal science. The
exploration of inner Asia, which was started only a century
ago in the wake of the Venetian's old experience, found a late
confirmation of its honest and durable value.

This is not an isolated case in the history of scientific
achievements. Empirical mechanics were instrumental in the
early development of experimental science. The great human
endeavors in establishing scientific truth were often preceded

by humble and candid attempts to satisfy intellectual curiosity. which is the most human of all instincts and the unquenchable impulse to all knowledge and learning.

HYPOTHESES NON FINGO

BY E. W. STRONG

Hitherto we have explained the phenomena of the heavens and of our sea by the power of gravity, but have not yet assigned the cause of this power. This is certain, that it must proceed from a cause that penetrates to the very centres of the sun and planets, without suffering the least diminution of its force; that operates not according to the quantity of the surfaces of the particles upon which it acts (as mechanical causes used to do), but according to the quantity of the solid matter which they contain, and propagates its virtue on all sides at immense distances, decreasing always as the inverse square of the distances. . . . But hitherto I have not been able to discover the cause of those properties from phenomena, and I frame no hypotheses *(Hypotheses non fingo);* for whatever is not deduced from the phenomena is to be called an hypothesis; and hypotheses, whether metaphysical or physical, whether of occult qualities or mechanical, have no place in experimental philosophy. In this philosophy particular propositions are inferred from the phenomena, and afterwards rendered general by induction. Thus it was that the impenetrability, the mobility, and the impulsive force of bodies, and the laws of motion and of gravitation, were discovered. And to us it is enough that gravity does really exist, and act according to the laws we have explained, and abundantly serves to account for all the motions of the celestial bodies, and of our sea *[Principia,* Book III, General Scholium]. [1]

To what does Newton have reference in saying that he frames no hypotheses? His declaration does not appear to fit his practice; for as Cajori notes, hypotheses were employed by Newton. [2] Cajori attempts to reconcile the apparent clash between profession and practice by arguing

. . . that Newton does not advance "hypotheses non fingo" as a general proposition, applying to all his scientific endeavor; it is used by him in connection with a public statement relating to that special, that difficult and subtle subject, the real nature of gravitation, which was mysterious then and has remained so to our own day. . . . Newton's "hypotheses non fingo" disrupted from its context is a complete misrepresentation of Newton.[3]

Cajori adds: "An examination of the various passages in Newton's writings, relating to the use of hypotheses, discloses the rule that experimental facts must invariably take precedence over any hypothesis in conflict with them. Secondly, hypotheses which seem incapable of verification by experiment are to be viewed with suspicion."[4]

To understand what Newton meant by *hypotheses non fingo*, " one must search out not only how Newton discourses about hypotheses but also ask what kind of hypothesizing Newton is talking about. Newton disavowed hypotheses earlier than the General Scholium, which was added to the second (1713) edition of the *Principia*, and he rejected them in connections other than with statements concerned with the real nature of gravitation. To see how Newton's ideas about hypotheses took shape, we shall attend to the occasions and course of his discussion on the subject, gleaning his sense as it is evident in what he says or as it can be gathered from the context.

On February 6, 1672, Newton sent a letter to Oldenburg for communication to the Royal Society. This letter "containing his New Theory about Light and Colours" announced the "doctrine" that "Light itself is a *Heterogeneous mixture of differently refrangible Rays*."[5] The evidence in support of this "doctrine" or "theory" consisted in a report of experiments which Newton had performed. He speculated that light was a substance ("whoever thought any quality to be a *heterogeneous* aggregate, such as Light is discovered to be"), and hypothesized its corpuscularian composition. "But, to determine more absolutely, what Light is, after what manner refracted, and by what modes or actions it produceth in our minds the Phantasms of Colours, is not so easie. And I shall not mix conjectures with certainties."

In separating the speculative conjecture from the theory founded upon experimental evidence, Newton is thereby insisting that the conjecture is not to be viewed as any essential part of his theory. The theory is delimited by setting aside all psychological suppositions about the production of mental affects as well as a more absolute determining of what light is. What, then, for Newton, is the proper determining and what certainty does it afford? As Newton proceeds in further communications to mark off *hypothesis* both from *theory* and *query*, what is intended in these distinctions made by Newton in defending his conclusions with respect to the method by which they were reached? He ends his first communication about his new theory with the assertion that it will itself "lead to divers experiments sufficient for its examination. " He thus indicates the procedure proper to "experimental philosophy. " Experiments are performed to discover the properties of things leading, through inductive generalization, to theory of the phenomena.

Robert Hooke reported on Newton's first paper at the meeting of the Royal Society held on February 15, 1672. In this report he contended that his own hypothesis would account as well as Newton's for the same phenomena "without any manner of difficulty or straining; nay, I will undertake to shew another hypothesis, differing from both his and mine, that shall do the same thing. "[6] Hooke says of Newton's theory that it is a hypothesis, but that he would not be understood to have criticized it for that reason; "for I do most readily agree with it in every part thereof, and esteem it very subtile and ingenious, and capable of solving all the phenomena of colours; but I cannot think it to be the only hypothesis, nor so certain as mathematical demonstrations. " Newton discussed both of Hooke's points in a letter to Oldenburg under date of July 11; but, by this time, he had already written two letters replying to objections raised against his new theory by Ignatius Gaston Pardies, a Jesuit father and a professor of mathematics in the Parisian College of Clermont.

Pardies first wrote on April 9, 1672, and Newton replied on April 13 in a letter to Oldenburg. Newton submits that the proper way to examine his new theory is by attending to the experimental evidence adduced in its support.

Now as to the Reverend Father calling my Theory *an Hypothesis*, I take it not amiss, because to him it may not yet appear. But I proposed it with another Intent, and it seems to contain nothing else but certain Properties of Light, which being now found out I think it not difficult to prove; and which if I did not know to be true, I would rather choose to reject as vain and empty speculation rather than to own them as my Hypothesis.[7]

Newton is not repudiating his conjecture about the corpuscularian composition of light. Then, and later, it seemed to him a likely explanatory supposition which could readily be "accommodated" to the phenomena. Rather, he is insisting that his experiments have discovered properties of rays of light which confirm his theory about their refrangibility. Matters of fact made evident by experiments are held to be independent of this, that, or the other hypothesis about light, e.g., whether it be corpuscularian, or a pression, pulse, or vibration of an ethereal medium. The conclusions "deduced" or "inferred" from experiments make up a theory inductively generalized from phenomena.

Pardies' second letter, dated May 21, 1672, brought up, as Hooke had already done, the question of alternative hypotheses. Both in his reply on June 2, 1672, to Pardies' second letter and in a letter to Oldenburg on July 8, 1672, Newton sets forth his reasons against trying to settle disputes about the properties of things solely by "a confutation of contrary suppositions." In the latter communication, he writes as follows:

Give me leave, Sir, to insinuate that I cannot think it effectual for determining truth, to examine the several waies by which phenomena may be explained, unless where there can be a perfect enumeration of all those waies. You know, the proper Method for inquiring after the properties of things is, to deduce them from Experiments. And I told you, that the Theory which I propounded, was evinced to me, not by inferring, *'tis thus because not otherwise,* that is, not by deducing it only from a confutation of contrary suppositions, but by deriving it from Experiments concluding positively and directly. The way therefore to examine it is, by considering whether the Experiments which I propound do prove those parts of the Theory, to which they are applied; or by prosecuting other Experiments which the Theory may suggest for its examination. And this I would have done in a due Method; the Laws of Refraction being thoroughly inquired into and determined before the nature of *Colours* be taken into consideration. It may not be amiss to proceed according to the *Series* of these *Queries;* which I could wish were determined by the Event of proper Experiments; declared by those that may have the curiosity to examin them.[8]

The eight queries which Newton proceeds to submit are hypotheses in question form. They are hypotheses in keeping with the following definition *(Oxford English Dictionary)*; hypothesis: "a supposition or conjecture put forth to account for known facts; especially in the sciences, a provisional supposition from which to draw conclusions that shall be in accordance with known facts, and which serves as a starting-point for further investigation by which it may be proved or disproved and the true theory arrived at." Newton, however, is not using the terms *query* and *hypothesis* interchangeably. Immediately following his expressed wish to have the queries determined by proper experiments, he writes:

To determine by Experiments these and such like Queries which involve the propounded Theory, seems the most proper and direct way to a conclusion. And therefore I could wish all objections were suspended, taken from Hypotheses or any other heads than these two: of shewing the insufficiency of Experiments to determine these Queries or prove any other parts of my Theory, by assigning the flaws and defects in my conclusions drawn from them; or of producing other Experiments which directly contradict me, if any such may seem to occur. For if the Experiments, which I urge, be defective, it cannot be difficult to show the defects; but if valid, then by proving the Theory, they must render all Objections invalid.[9]

The specification for the framing and use of a *query* (an interrogative hypothesis) can now be laid down in keeping with the foregoing quotations from Newton:

1. A query conjectures concerning a proposed theory by asking whether something pertinent or crucial to the theory (or some part of it) is or is not so in fact.

2. The reply to a query, and thus the evidential testing of the theory, rests on the sufficiency or adequacy of experiments to determine what is so.

3. A query is framed with reference to its correspondence with facts in the double sense (a) that the theory involved in framing the query is (for Newton) an inductive generalization from known facts and hence a body of facts lies behind the questioning pertaining to the theory, and (b) the query is framed to settle some doubt about the theory put to the test of experiments.

4. A query involving a propounded theory, when determined by experiments to which no exceptions are taken and against which no contrary experimental evidence is submitted, proves the theory (or some portion of it) subject to any further queries which the theory may suggest for its examination or to any additional experimental findings.

The specifications listed are those of a hypothesis used to test a theory propounded to account for some body of known facts; or, put another way, that which is hypothesized in a theory is brought to the fore and subjected to a test by framing leading questions to be settled by experiments. Under the name of *queries,* then, Newton not only frames hypotheses but discourses perspicaciously about their methodological use in physical science. In approving *queries* while, at the same time, viewing hypotheses with suspicion mounting to impatience, Newton is either muddled in his thinking or he has in mind a meaning of *hypothesis* quite different from the kind of hypothesizing he has included under *queries*.

The meaning which Newton attaches to "explaining by hypotheses" emerges in the course of further communications from him about his theory of light and colors. Pardies, in his second letter, [10] pointed out how the hypothesis of Grimaldi or of Hooke might serve to explain certain of the phenomena reported by Newton. Newton in reply makes clear that he does not find this argument concerning alternative hypotheses to be a relevant objection to his theory.

That I may answer this, it is to be observed that the doctrine I have explained concerning refraction and colours, consists only in certain properties of light, neglecting all hypotheses by which those properties may be explained. For the best and safest way of philosophizing seems to be, first, to enquire diligently into the properties of things, and to establish these by experiments, and to proceed later to hypotheses for the explanation of things themselves. For hypotheses ought to be applied only in explanation of properties of things, and not made use of in determining them; except in so far as they may furnish experiments. And if anyone offers conjectures about the truth of things from the mere possibility of hypotheses, I do not see by what stipulation any thing certain can be determined in any science; since one or another set of hypotheses may always be devised which will appear to supply new difficulties. Hence I judged that one should abstain from contemplating hypotheses as from improper argumentation, and that the force of their opposition must be removed, that one may arrive at a maturer and more general explanation. [11]

To insist on finding out the properties of things by experiments before proceeding to hypotheses for their explanation is to insist on an empirical grounding of suppositions. A hypothesis, when so grounded, has plausibility (i.e., some likelihood or probability) and in this is to be contrasted with merely possible or merely speculative conjectures. Newton's earlier contention that properties of things established by experiments are independent of one or another hypothesis is reiterated in holding that hypotheses are not to be used to determine properties "except in so far as they may furnish experiments." So used, there is no significant difference between a grounded hypothesis and a query. As queries involving a theory are settled, the explanation which the theory tenders is proved or disproved. Newton's rejection of "explanation by hypotheses" pertains, then, not to suppositions so used or applied, but rather to conjecturing about the truth of things "from mere possibility of hypotheses." He does not see "how any thing certain can be determined in any science" by such conjecturing.

Newton emphatically declares that his theory was evinced to him not by deducing it only "from a confutation of contrary suppositions, but by deriving it from Experiments concluding positively and directly." Unless there were a "perfect enumeration" of "contrary possibilities," the inference, "tis thus because not otherwise," would remain inconclusive. Perfect enumeration is an ideal condition, requiring for its satisfaction that all possibilities have been exhausted and hence that the true hypothesis has been included. To have the possibilities be contrary, in the sense of mutually exclusive, is also an ideal condition. If hypotheses overlap, or if each of two or more hypotheses accommodates all known facts in explanation of them, they are not contrary in the sense of only one true and all others false. Nor is this the sum of uncertainty which attends the attempt to come to conclusion in explanation (or to the truth of things) by sifting of mere possibilities, since further hypotheses may always be contrived without assurance

that these are free from new difficulties in place of those which
an investigator is seeking to overcome.

All in all, Newton rejects for explicit reasons and in the interest of positive knowledge in science what he judges to be
a wrong way--a way of fallacious argumentation entered upon
by those who would rest explanations upon the "possibility" of
hypotheses. Had he said only that he thought it proper to "lay
aside," "abstain from," and "decline" all hypotheses or hypothesizing *of this kind,* no puzzle would arise here; nor, since
such asseveration is one with saying *hypotheses non fingo,*
would there have been reason to wonder about his meaning
later. What he does say is that he thinks it proper for the examination of his theory by experiments to lay aside all hypotheses. [12] The assertion is too sweeping because it includes the
grounded hypothesizing embraced under *queries* which Newton
explicitly advises. There is no ambiguity, however, about determining properties of things merely from a confutation of
contrary suppositions. Such procedure is rejected. A theory
concluded from experiments is to be tested by experiments.

Turning to the letter to Oldenburg of July 11, 1672, in which
Newton replies to Hooke's review of his new theory, we find
Newton acknowledging that he has argued the corporeity of
light from his theory:

> . . . but I do it without absolute positiveness, as the word *per-
> haps* indicates, and make at most but a very plausible consequence
> of the doctrine, and not a fundamental supposition, nor so much as
> any part of it, which was wholly comprehended in the precedent prop-
> ositions. And I wonder how Mr. Hook could imagine, that when I
> had asserted the theory with the greatest rigor, I should be so for-
> getful, as afterwards to assert the fundamental supposition itself
> with no more than a *perhaps*. [13]

So far, Newton differentiates between asserting his corpuscularian speculation tentatively as at most "a very plausible
consequence" of his theory, and treating it as a foundation upon
which his theory stands or falls.

Hooke had said of Newton's new theory, "I cannot think it to
be the only hypothesis, nor so certain as mathematical demonstration." A reply to the first part of the objection is given in
Newton's assertion that he has not argued to the exclusion of

other hypotheses in venturing his own, and that he had not intended or proposed dependence of his theory upon the corpuscularian hypothesis.

Had I intended such *hypothesis*, I should somewhere have explained it. But I knew that the *properties*, which I declared of light, were in some measure capable of being explicated not only by that, but by many other mechanical hypotheses; and therefore I chose to decline them all, and speak of light in general terms, considering it abstractly as something or other propagated every way in streight lines from luminous bodies, without determining what that thing is; whether a confused mixture of difform qualities, or modes of bodies, or of bodies themselves; or of any virtues, powers or beings whatsoever. [14]

Newton has now made clear, if he has not before, what kind of hypothesizing he chooses to avoid. Instead of framing some fundamental supposition about what light is (later, in the *Principia*, about what gravity is) and laying down this supposition as a premise of theory about properties, he leaves such ultimate causes to be found out. It suffices, for his avowed purpose, to ground his theory on experimental evidence, settling disputes about theory, if any arise, by framing leading questions crucial to it to be determined by experiments. Newton reiterates this position set forth in his reply to Hooke when, in April, 1673, in reply to objections to his theory raised by Christian Huygens, he declares that

. . . to examine, how Colours may be explain'd *hypothetically*, is beside my purpose. I never intended to shew, wherein consists the Nature and Difference of Colours, but only to shew, that *de facto* they are the Original and Immutable qualities of the Rays which exhibit them; and to leave it to others to explicate by Mechanical *Hypotheses* the Nature and Difference of those qualities. . . . [15]

A like declaration is made in a letter from Newton to Oldenburg dated at Cambridge, February 15, 1675/6. Protesting against those who raise objections by propounding "an Hypothesis to explain my Theory," Newton writes:

I shall now for a conclusion remind you of what I have formerly said in general to the same purpose so that I may at once cast off all objections that may be raised in the future either from this or any other Hypothesis whatever. If you consider what I said both in my second letter to P. Pardies & in my answer to Mr. Hook, Sect. 4. concerning the application of all Hypothesis to my Theory you may thence gather this general rule: *That in my Hypothesis where the rays are supposed to have any original diversities, whether as to size or figure or motion or force or quality or anything else imaginable which may suffice to difference those rays in colour and refrangibility, there is no need to seek for other causes of these effects than those original diversities.* This rule being laid down, I argue thus. In my Hypothesis, whether light as it comes from the sun may be supposed either homogeneal or heterogeneal. If the last, then is that Hypothesis comprehended in the general rule and so cannot be against me; if the first then must refraction have a power to modify light so as to change its colorifick qualification and refrangibility; which is against experience.[16]

With respect to Newton's proposal to "abstract the difficulties involved in Mr. Hook's discourse" in order to meet these "without regard to any hypothesis," Cajori observes that Newton dismisses "the two major hypotheses of light as wave motion and as flying particles, in order to consider three minor hypotheses, more easily disposed of by experimental test, but nevertheless hypotheses, although he does not designate them by that name."[17] The name by which Newton does designate them is *queries.* Newton, then, is being consistent with his earlier expressed concern to admit hypotheses only so far as they furnish experiments testing them, and consistent in the usage by which he allocated to the name "query" the role in science now designated by "hypothesis." The fact that Newton turned his consideration to the three suppositions most readily testable by experiments accords with the profession made by him about the proper method to be followed in experimental philosophy. Had he seen how the two major hypotheses could be brought to experimental testing, there is no reason to think he would have asked to be excused from taking them up or thought it beside the business at hand to consider them. It is the disputing about hypotheses in the absence or lack of testing by experiments which Newton suspects and declines.

Reply to Hooke's second point, that Newton's theory is not so certain as mathematical demonstration, is given in the following words:

I said, indeed, that the science of colours was mathematical, and as certain as any other part of Optics; but who knows not that Optics, and many other mathematical sciences, depend as well on physical sciences, as on mathematical demonstrations? And the absolute certainty of a science cannot exceed the certainty of its principles. Now the evidence, by which I asserted the propositions of colours, is in the next words expressed to be from experiments, and so but physical: whence the Propositions themselves can be esteemed no more than physical principles of a science. [18]

No theory formulated by inductive generalization will be so certain as to exempt it from further questioning; for induction cannot insure but only assume continued constancies of processes and structures of nature. If an issue crucial to a theory arises, "Tis better," Newton holds, "to put the Event on further Circumstances of the Experiment, than to acquiesce in the Possibility of any Hypothetical Explications."[19] Experimental testing of theory is the proper task that must not be neglected in a physical science. Merely to speculate and to dispute about conjectured natures and causes is prejudicial to conclusiveness. The possibility of hypothetical explications should not be substituted for the determination of queries by experiments.

When, in the General Scholium added to Book III of the *Principia* in 1713, Newton defines hypothesis as "whatever is not deduced from phenomena," hypothesis so defined is a merely speculative supposition. Yet it is evident from passages so far cited that Newton's reluctance to frame hypotheses extends to suppositions accommodating known facts and laws. Although these, unlike merely speculative suppositions, have some initial probability, the more they are lacking in experimental verification the more Newton desires to lay them aside. The bent of his mind in this respect is revealed in his letter to Robert Boyle (February 28, 1678/9)[20] in which he conjectures about the cause of gravity. He first supposes diffusion through all places of an "aetherial" substance capable of contraction by the graduated subtility of the ether. If, says Newton, there is "any degree of probability" in these conjectures, this is all he aims at. "For my own part, I have so little fancy to things of this nature, that, had not your encouragement moved me to it, I should never, I think, have thus far set pen to paper about them."

Even with a sympathetic audience and no unwelcome controversy to vex him, Newton was averse to the kind of hypothesizing in which he had, on this occasion, indulged. In the setting of controversies over his theory of light and colors, Newton's native caution was irked and increased by the way in which objectors seemed to him to run to uncertain speculations when they ought rather to have stayed with the experimental evidence. Thus of Hooke's criticism he writes in his letter of July 11, 1672, "But, I must confess at first receipt of those considerations, I was a little troubled to find a person so much concerned for an *hypothesis*, from whom in particular I most expected an unconcerned and indifferent examination of what I propounded."[21] Graciously, he adds, "But yet I doubt not we have one common design; a sincere endeavour after knowledge, without valuing uncertain speculations for their subtleties, or despising certainties for their plainness."

The tone is more curt in 1675 when, on December 21, Newton writes to Oldenburg at the latter's insistence to reply to "Mr. Hook's insinuation, that the sum of the hypothesis I sent you, has been delivered by him in his Micography."[22] Newton first indicates what he takes to be Hook's extensive borrowing from Descartes and others in asserting that there is an ethereal medium and that light is the action of this medium. Having summed up Hooke's modification of the Cartesian hypothesis, Newton declares, "in all this, I have nothing in common with him but the supposition that the ether is a medium susceptible of vibrations. Of which supposition I make a very different use; he supposing it light itself; which I suppose it is not." The sum of what is common to Hooke and himself reduces, Newton says, to the assumption that the ether may vibrate. While acknowledging use made of Hooke's observations, Newton adds, "But he left me to find out and make such experiments about it, as might inform me of the manner of those colours to ground an hypothesis on."[23]

Newton, in the sentence just quoted, takes cognizance of hypothesis grounded on experiments. This is not, however, the way in which he characterizes hypotheses in the *Principia* and the *Opticks*. Hypotheses are there designated as suppositions not deduced from phenomena. Hypothetical explanations are by assumed causes unverified by experiments. They are, thus, not to be made part of the conclusions drawn by induction from phenomena in formulating general propositions of a physical theory. Reasoning in the demonstrations of theory proceeds from causes discovered and established as "prin-

ciples" or "axioms" by which the phenomena are explained
and the explanations proved. Newton's scientific purpose, as
announced near the beginning of *The System of the World*,

> . . . is only to trace out the quantity and properties of this force
> (gravitation) from the phenomena, and to apply what we discover
> in some simple cases as principles, by which, in a mathematical
> way, we may estimate the effects thereof in more involved cases,
> for it would be endless and impossible to bring every particular to
> direct observation. We said, *in a mathematical way,* to avoid all
> questions about the nature or quality of this force, which we would
> not be understood to determine by any hypothesis. . . .[24]

The methodological reason for not conjecturing about the
cause of gravity is supplied in Newton's fourth rule of reasoning in philosophy.

> In experimental philosophy we are to look upon propositions inferred
> by general induction from phenomena as accurately or very nearly
> true, notwithstanding any contrary hypotheses that may be imagined,
> till such time as other phenomena occur, by which they may either
> be made more accurate, or liable to exception. This rule we must
> follow, that the argument of induction may not be evaded by Hypoth-
> eses.[25]

The same expression of scientific purpose appears in the
Opticks (1704), where, instead of properties of gravity, it is
properties of light which Newton does not intend to determine
or to explain by hypotheses: "My design in this Book is not to
explain the Properties of Light by Hypotheses, but to propose
and prove them by reason and experiments."[26] The propositions about optical phenomena, upon which investigators are
generally agreed, he assumes under the name of "axioms"
or "principles" in explanations and demonstrations of phenomena.

Appended to the work in the first edition were sixteen "queries" to which Newton added seven more in the Latin edition
(Optice, 1706) and eight more in the second English edition of
1718. Unlike Newton's earlier use and specification of queries as involving a theory and leading to experiments determining them and thus testing the theory, these range all the
way from such a restricted role to speculations in natural the-
ology about an intelligent Creator of nature's order. While af-
firming his faith in design by an "Intelligent Agent," Newton

at the same time makes clear that the truth of general laws of physical phenomena is not made to rest within his scientific accounting upon conjectures about a cause, or causes, of natural laws, but derives from experimental evidence handled in a mathematical way. "And therefore I scruple not to propose the Principles of Motion above-mention'd, they being of a very general Extent, and leave their Causes to be found out."[27] The position expressed in these words from query 31 is the same voiced in the General Scholium. Newton avers that, having explained the phenomena of the heavens and of our sea by the power of gravity, he has "not yet assigned the cause of this power." No hypothesis is framed for the reason that he has not hitherto "been able to discover the cause of these properties from the phenomena." Or, as Newton put it in query 28, "the main business of natural philosophy is to argue from Phaenomena without feigning Hypotheses. . . ."[28]

In saying *"Hypotheses non fingo,"* Newton is asserting his intention to avoid suppositions not deduced from phenomena, whether the hypotheses are metaphysical or physical and whether of occult qualities or mechanical. He repeats this declaration in the *Opticks* in the following sentences added in 1718 to query 31:

For Hypotheses are not to be regarded in Experimental Philosophy. And although the arguing from Experiments and Observations by Induction be no Demonstration of general Conclusions; yet it is the best way of arguing which the Nature of Things admits of, and may be looked upon as so much the stronger, by how much the Induction is the more general. And if no Exception occur from Phaenomena, the Conclusion may be pronounced generally. But if at any time afterwards any Exception shall occur from Experiments, it may then begin to be pronounced with such Exceptions as occur.[29]

For Newton, theoretical conclusions stand in the middle of evidence in being propositions derived inductively from experiments and subject to exceptions which may arise in further experiments. Under the name of "queries," grounded hypothesizing is recognized, advised, and employed. The role of such hypothesizing and the method by which it should be employed are set forth explicitly. Queries involving a theory ask leading questions for the test of the theory in determination of the queries by experiments. The proper course to follow in objecting to a theory is either to show that the evidence so far submitted is inadequate to sustain conclusions drawn from it or to produce other experiments running counter to the theory. It is another kind of hypothesizing which Newton intends to exclude as not part of his purpose or design in natural philosophy. The rejected hypothesizing speculates about possible causes and natures of things, instituting such possible explanations as fundamental suppositions upon which the truth of a theory is made to depend. Newton's reasons for rejecting hypotheses of this sort in this role are, in summary, the following:

1. Propositions inferred by general induction from phenomena are to be regarded as accurately or very nearly true prior to, and independent of, satisfying in practice the ideal conditions of perfect enumeration and of mutual exclusiveness of "contrary suppositions."

2. Attempt to settle disputes about the properties of things and the truth of explanations solely by "confutation of contrary suppositions" or by "mere possibility of hypotheses" is not only prejudicial to certainty in science but also, in proposal of one or another further hypothesis to overcome preceding difficulties, attended with new tribulations.

3. Disputing about alternative possible explanations (unless these furnish experiments) is not the proper way to advance knowledge in establishing the truth of propositions, for the proper method reasons from experiments and observations in establishing positive knowledge and such argument of induction is evaded when merely speculative suppositions are allowed to stand in the way of coming to a conclusion.

No wonder, then, that Newton wrote, *"Hypotheses non fingo."*

THE FIRST STELLAR
PARALLAX DETERMINATION

BY OTTO STRUVE

Two Epochs in the History of Astronomy

At the Rome meeting of the International Astronomical Union in September, 1952, Walter Baade of the Mount Wilson and Palomar Observatories electrified the entire astronomical world by his announcement of a new distance scale of the universe which exceeded the previously accepted scale by a factor of two. The most distant galaxies observed at Mount Wilson are one billion light years away, instead of five hundred million; and the penetrating power of the two-hundred-inch telescope is actually twice as great as had been anticipated.[1]

This tremendous advance in our knowledge of the geometry of the universe has come about, in part, through the slow accumulation of accurate observations by several astronomers, and, in part, through the skillful use by Baade and his associates of some of the first photographs obtained with the two-hundred-inch telescope. The exact history of the events that culminated in the Rome announcement is clear to all of us who have witnessed them. They will remain clear to posterity only if this history is now recorded in complete detail, with fairness and with a thorough evaluation of the various contributing elements.

Our present object is not to write this history. But it is appropriate to mention the events because they remind us of the history of another, even greater, epoch in the history of man's knowledge of the geometry of the universe--an epoch which cul-

minated in the years 1837, 1838, and 1839, when the first reliable distances of three fixed stars were announced. As was
the case in the more recent advance, the true distance scale
of the system of the fixed stars was then still in doubt; and the
similarity of these two astronomical revolutions, the one of
1952 and the other 115 years earlier, is such as to involve in
both cases a great increase in the scale--by a factor of two
in 1952; by a factor of approximately ten in 1837-39.

How great an importance was attached by contemporary astronomers to this earlier advance is well stated in Sir J. F.
W. Herschel's address before the Royal Astronomical Society
of London, on presenting to the Königsberg astronomer, F. W.
Bessel, the gold medal of the society for his measurement of
the distance of the double star 61 Cygni:

. . . I congratulate you and myself that we have lived to see the
great and hitherto impassable barrier to our excursions into the
sidereal universe; that barrier against which we have chafed so
long and so vainly--*(aestuantes angusto limite mundi)*--almost simultaneously overleaped at three different points. It is the greatest
and most glorious triumph which practical astronomy has ever witnessed. Perhaps I ought not to speak so strongly--perhaps I should
hold some reserve in favour of the bare possibility that it may be all
an illusion--and that further researches, as they have repeatedly
before, so may now fail to substantiate this noble result. But I confess myself unequal to such prudence under such excitement. Let
us rather accept the joyful omens of the time, and trust that, as
the barrier has begun to yield, it will speedily be effectually pros-
trated. Such results are among the fairest flowers of civilisation.
They justify the vast expenditure of time and talent which have led
up to them; they justify the language which men of science hold, or
ought to hold, when they appeal to the governments of their respective countries for the liberal devotion of the national means in furtherance of the great objects they propose to accomplish. They enable them not only to hold out but to redeem their promises, when
they profess themselves productive labourers in a higher and richer
field than that of mere material and physical advantages. It is then
when they become (if I may venture on such a figure without ir-
reverence) the messengers from heaven to earth of such stupendous
announcements as must strike every one who hears them with almost awful admiration, that they may claim to be listened to when
they repeat in every variety of urgent instance, that these are not
the last of such announcements which they shall have to communi-
cate, --that there are yet behind, to search out and to declare, not

only secrets of nature which shall increase the wealth or power of man, but TRUTHS which shall ennoble the age and the country in which they are divulged, and by dilating the intellect, react on the moral character of mankind. Such truths are things quite as worthy of struggles and sacrifices as many of the objects for which nations contend, and exhaust their physical and moral energies and resources. They are gems of real and durable glory in the diadems of princes, and conquests which, while they leave no tears behind them, continue for ever unalienable. [2]

This medal was then regarded--and it still retains this distinction today--as the greatest recognition that an astronomer can hope to achieve. John Herschel's account of the circumstances under which the "hitherto impassable barrier" was penetrated is an accurate and eminently fair appraisal of the contributions that finally brought success; but, of course, this appraisal was made by astronomers who knew only what had preceded the great work of Bessel; and in the passage quoted above Sir John admits the remote possibility "that it may be all an illusion."

Is there now any reason to question John Herschel's account and to rewrite the history of the first parallax determination? This question has been put forward in a recent article by A. N. Deitch, astronomer of the Pulkovo Observatory in Russia.[3] I quote from his introductory paragraph:

It is well known that the first reliable measurements of the distances of the stars were carried out almost simultaneously by three astronomers: W. Struve, Bessel and Henderson. However, Struve's contributions in this field are not accorded their proper recognition, and his determination of the parallax of Vega is often described incorrectly. For example in A. Clerke's book "History of Astronomy in the XIX-th Century" (translated into Russian by V. V. Serafimov, in 1913) we read on page 60: 'W. Struve, then already at Pulkovo, and making use of the new 15-inch refractor, devoted his attention to the brilliant gem in the constellation Lyra, whose Arabic name is Vega . . . "·In reality, W. Struve, observed the parallax of Vega with the 9-inch refractor in Dorpat (Yuriev, now Tartu).[4] The same error occurs in A. Berry's short history of astronomy (edited, with additions by P. V. Kunitzky, 2nd edition, 1946) on page 308: ". . . W. J. Struve obtained for the star Vega a parallax of 1/4 second of arc, from observations made at Pulkovo."[5] The impression has gained ground that the parallax of α Lyrae was published by Struve

a year later than the investigations by Bessel and Henderson, and
that the value of this parallax was in greater discord with modern
data (see for example, P. P. Parenago: "Textbook of Stellar Astronomy, "[6] 2nd edition 1946, page 31).

It would appear that Dr. Deitch's comments involve two distinct points: (1) he calls attention to a considérable amount of
confusion in the current astronomical literature with regard
to the chronology of the events, and (2) he believes that the
work of W. Struve has not been accorded its proper amount of
recognition.

Neither of these two points is entirely new. The second has
been ably discussed by one of the leading living astronomers--
Dr. J. Jackson, formerly His Majesty's Astronomer at the
Cape of Good Hope Observatory, and at the time of his histori-
cal discussion of stellar parallaxes, chief assistant of the
Greenwich Observatory. I quote from his paper:

> The arguments given show how keenly astronomers a century
> ago felt the need for the determination of the parallax of a star and
> how brilliantly they tackled the problem. It will be seen how system-
> atically every possible avenue of approach was examined, and as
> we read the pages of Struve we feel that the writer was certain that
> the solution was near, and that it was just within the reach of micro-
> metric observation with his telescope. The hopes of many genera-
> tions had remained unrealised, but Struve was not despondent--he
> almost was expectant when at last success did come. We have said
> little of what Bessel did, although he did much to improve the meth-
> ods of reduction and generally inspired confidence in his results. He
> also devised and applied the heliometer. To him is usually assigned
> the chief merit for the actual determination of the parallax of a star.
> The name of Henderson is generally added for his observations of
> α Centauri. It is only sometimes that the name of Struve is linked
> with the final success. It is hoped that this article shows more clear-
> ly the greatness of the contributions made by Struve, both from the
> theoretical and from the observational point of view.[7]

The first point of Deitch's remarks was already briefly
mentioned in my review of a wartime volume by W. M. Smart,
entitled *Foundations of Astronomy*.[8] In this review, I expressed my opinion concerning the questions raised by Deitch:

> On page 128 Smart briefly refers to the history of parallax determi-
> nations for fixed stars and repeats an historical error which has
> crept into the newer astronomical literature, though it contradicts

Fig. 1. View of the Dorpat Observatory as it appeared in Struve's time

Fig. 2. View of the Pulkovo Observatory the construction of which was Struve's chief concern in the 1830's. The observatory was dedicated in 1839, and was totally destroyed by the Germans during the last war. It has been rebuilt by the Soviet government. (Photograph from *Zum 50-Jährigen Bestehen der Nikolai-Hauptsternwarte* by O. Struve [St. Petersburg, 1889])

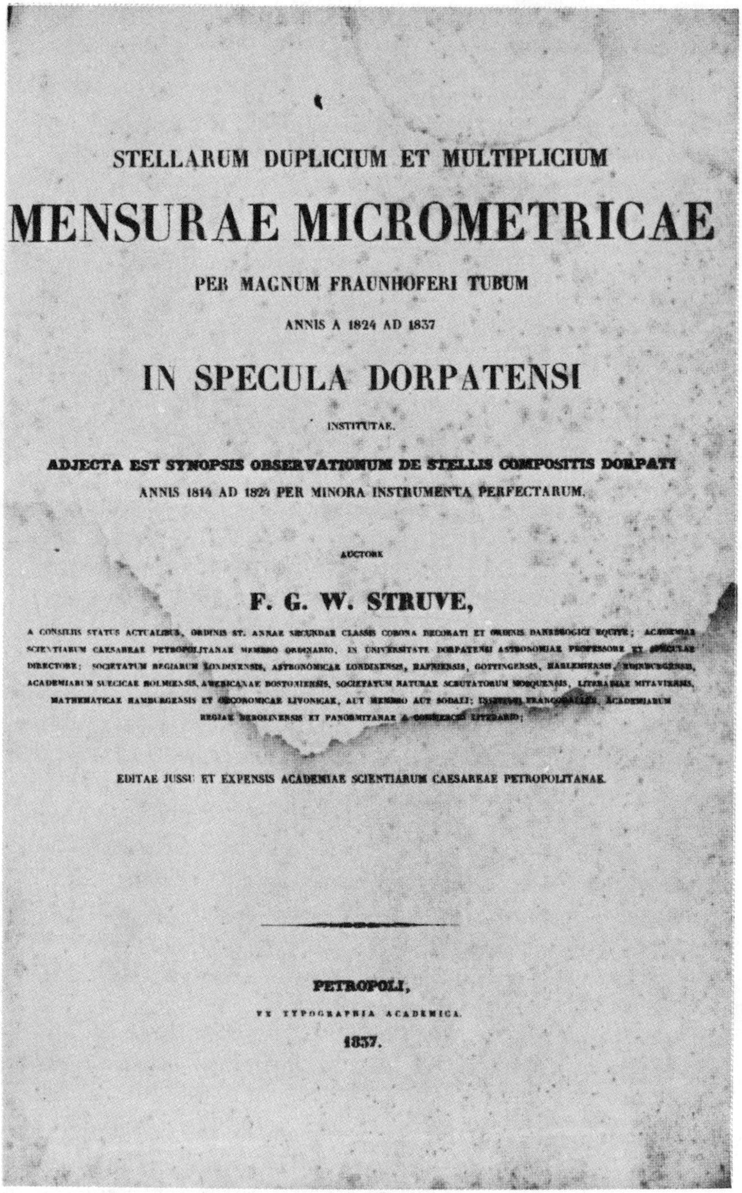

STELLARUM DUPLICIUM ET MULTIPLICIUM

MENSURAE MICROMETRICAE

PER MAGNUM FRAUNHOFERI TUBUM

ANNIS A 1824 AD 1837

IN SPECULA DORPATENSI

INSTITUTAE.

ADJECTA EST SYNOPSIS OBSERVATIONUM DE STELLIS COMPOSITIS DORPATI
ANNIS 1814 AD 1824 PER MINORA INSTRUMENTA PERFECTARUM.

AUCTORE

F. G. W. STRUVE,

A CONSILIIS STATUS ACTUALIBUS, ORDINIS ST. ANNAE SECUNDAE CLASSIS CORONA DECORATI ET ORDINIS DANEBROGICI EQUITE; ACADEMIAE
SCIENTIARUM CAESAREAE PETROPOLITANAE MEMBRO ORDINARIO, IN UNIVERSITATE DORPATENSI ASTRONOMIAE PROFESSORE ET SPECULAE
DIRECTORE; SOCIETATUM REGIARUM LONDINENSIS, ASTRONOMICAE LONDINENSIS, HAFNIENSIS, GOTTINGENSIS, HARLEMENSIS, MONACENSIS,
ACADEMIARUM SUECICAE HOLMIENSIS, AMERICANAE BOSTONIENSIS, SOCIETATUM NATURAE SCRUTATORUM MOSQUENSIS, LITERARIAE MITAVIENSIS,
MATHEMATICAE HAMBURGENSIS ET OECONOMICAE LIVONICAE, AUT MEMBRO AUT SODALI; INSTITUTI FRANCOGALLICI, ACADEMIARUM
REGIAE BEROLINENSIS ET PANORMITANAE A COMMERCIO LITERARIO;

EDITAE JUSSU ET EXPENSIS ACADEMIAE SCIENTIARUM CAESAREAE PETROPOLITANAE.

PETROPOLI,
EX TYPOGRAPHIA ACADEMICA.
1837.

Fig. 3. Title page of the *Mensurae Micrometricae*

some of the earlier accounts. He states that the first successful measurement of a stellar parallax was that of 61 Cygni in 1838, which was quickly followed by the measurements of α Centauri and α Lyrae. As a matter of fact, the first reliable parallax was that of α Lyrae published in 1837 in the Latin Introduction to the *Mensurae Micrometricae* (by F. G. W. Struve). The value there given is $\pi = 0''.125 \pm 0''.055$. The best modern value is $\pi = 0''.121 \pm 0''.004$. The first reliable parallax of 61 Cygni was published in 1838 and was actually obtained by Bessel in that year. The important thing, however, is not, as Smart's statement might lead one to think, which parallax was determined first, but which parallax actually dispelled all doubts of the contemporary astronomers that the long-searched-for effect had finally been found. There can be no doubt that the parallax of 61 Cygni, and not that of α Lyrae, gave this assurance: it had a probable error of only $\pm 0''.014$ applied to a parallax of $\pi = +0''.314$. The best modern value is $\pi = 0''.299 \pm 0''.003$. The relatively much greater uncertainty of the first parallax of α Lyrae is also demonstrated by the fact that later micrometer observations (also by F. G. W. Struve), up to 1838, gave $\pi = 0''.261 \pm 0''.025$, while observations with a vertical circle by Peters gave, in 1846, $\pi = 0''.103 \pm 0''.053$. Bessel's additional observations of 61 Cygni, published in 1840, gave $\pi = 0''.348 \pm 0''.010$.[9]

I believe it is important to distinguish which result appeared convincing to the contemporaries of Bessel, Struve, and Henderson; and what we, with the advantage of 115 years of subsequent investigations may be able to recognize as the first successful penetration of Herschel's "barrier." The astronomers of the early years of the nineteenth century were understandably skeptical of new announcements of stellar parallaxes, after the efforts of scores of the best observers from Tycho Brahe to Bradley had failed. It should also be remembered that both Bessel and Struve had been occupied with parallax measurements--all unsuccessful--long before they triumphed in 1837 and 1838.

That there is much confusion concerning the first parallax determination is already apparent from the comments by Deitch. It is easy to add to the specific instances referred to in his article. We shall list only a few chosen almost at random. R. Wolf, whose history of astronomy is one of the standard works on this subject, writes:

Soon after Bessel, Wilhelm Struve also undertook . . . a similar determination for which he chose α Lyrae--one of the brightest and presumably closest stars--and communicated in the 1839 addition to his "Mensurae Micrometricae" entitled "Disquisitio de Parallaxi α Lyrae" that he had obtained a parallax of 0''.26 or a distance of 16 billion miles. [10]

We have already seen that the first announcement appeared in 1837 in the *Mensurae Micrometricae* themselves and was there given as $\pi = 0''.125$. Wolf mentions the name Henderson only casually: "Since that time similar determinations have been made by Thomas Henderson, John Brinkley, Otto Struve, Peters. . . ." In reality, Henderson's observations of α Centauri at the Cape preceded, in part, those of Bessel and Struve, but he delayed the discussion and publication of his results until 1839. [11]

F. Becker[12] assigns all three discoveries to 1838: "Almost simultaneously Bessel succeeded in measuring the parallax of 61 Cygni, W. Struve that of Vega, and Henderson that of Alpha Centauri. " E. Lebon writes: "Bessel found in 1838 the parallax of 61 Cygni thanks to the excellent heliometer constructed in 1829 by Fraunhofer. W. Struve, who was the first to adapt the method of Galileo to practice, obtained in 1840 the parallax of Alpha Lyrae. "[13] Henderson is not mentioned at all. In reality, Bessel and Struve both used the method suggested by Galileo--of measuring the differential parallax of two (or more) neighboring stars, one of which is believed to be much nearer to us than the other. They differed only in the form of the instrument: Bessel used the micrometer attached to a split heliometer objective; Struve used a filar micrometer in the focal plane of the telescope.

W. Brunner[14] mentions only Bessel. P. Couderc attributes all three parallax determinations to 1840 and assigns to Struve the parallax of 0''.1 for Vega. [15] We have seen that this was the value found in 1837. Struve's later determination, published in 1839, was 0''.261. E. A. Fath assigns the three determinations to 1838 and then remarks: "Bessel, Struve and their successors used an instrument called a heliometer. "[16] In reality the heliometer was used by Bessel, while Struve used a filar micrometer. E. Zinner[17] also attributes the first determination to Bessel in 1838 and mentions for α Lyrae the later value of 0''.26, and for α Centauri Henderson's result of 0''.92. R. L. Waterfield[18] discusses only the parallaxes of 61 Cygni and α

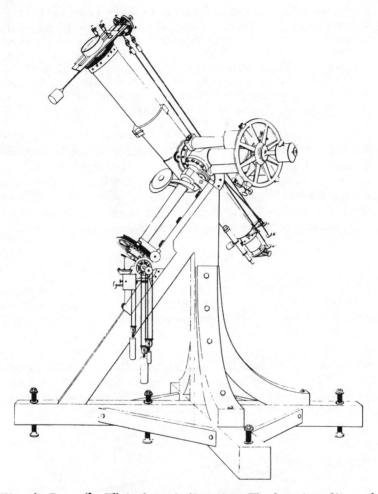

Fig. 4. Bessel's Königsberg heliometer. The lens is split, and the two halves can be moved with respect to one another with the help of a micrometer screw. The aperture is 6 1/4 inches and the focal length is 102 inches. (Plate XV from *Astronomische Nachrichten*, VIII [1830], 398)

Cenaturi, and so does R. A. Sampson who says with regard to the latter:

Henderson has sometimes been blamed for undue caution and delay.
This seems a wrong view of the case; with the means at his disposal,
caution and confirmation were an obligation. After his results were
confirmed, the Council felt that he, too, should have recognition.
But they missed the right opportunity for action. In 1843 the material
was before them, and no name proposed for the medal. In 1844,
November, Henderson's name was put forward, but in the same
month he died. [19]

The last (eighth) German edition of Newcomb-Engelmann's
Populäre Astronomie (1948)[20] attributes the first two determinations to Bessel ("one had here, for the first time, determined the distance of a fixed star by a method that deserved
full confidence") and to Struve, whose second parallax (of
1839) is given for Vega. Henderson's result is mentioned as
"approximately at the same time as Bessel and Struve."

The Trigonometric Method

The earliest method used by astronomers in their effort to
measure the distances of the fixed stars is illustrated in Figure 5.* Two stars (S_1, S_2) were chosen such that they made
an angle of 180° when the earth was in E_1. Half a year later,
with the earth in E_2, the angle should then be less than 180°
and the departure from 180° will be twice the sum of the parallaxes, $2(\pi_1 + \pi_2)$. When the angle is, to begin with, not exactly 180°, and when the stars are not in the ecliptic (the plane
of the earth's orbit), the result is slightly more complicated,
but the main idea remains the same.

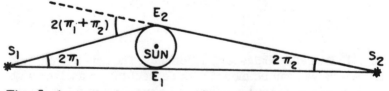

Fig. 5. An early, but unsuccessful, method of measuring the
distances of fixed stars

---

*The author's diagrams in Figures 5-8 and 11 first appeared
in *Sky and Telescope*, XVI (1956), 9, 69.

Since it is not possible to measure directly angles of the order of 180°, this method requires the determination of the absolute coordinates of the two stars--a procedure that involves many sources of error. Nevertheless, this method was already used by Copernicus. From the absence of any detectable effect he concluded that the stars must be at least one thousand times more distant than the sun. With more refined instruments Tycho Brahe also failed: there was no measurable angle $(\pi_1 + \pi_2)$. Hence the stars had to be either more than three thousand times as far as the sun, or the earth did not revolve around the sun. Tycho Brahe chose the second alternative: to him it appeared inconceivable that the stars should be at such enormous distances.

Later investigations, up to and including the earlier work of Bessel and Struve, made frequent use of this method, but the results always were negative: there was no measurable angle $(\pi_1 + \pi_2)$.

However, another method had already been suggested, independently, by Galileo and by Huygens. It was to measure the difference between the parallactic displacements of two stars that are close to each other in the sky but are, in reality, at different distances from the sun. Figure 6 illustrates this effect. The orbital motion of the earth causes the stars to describe small ellipses on the celestial sphere. These ellipses are large in angular size when the stars are near to us, and they become vanishingly small when the stars are very distant. We could, for example, measure the angular distance between the two stars in Figure 6 and from their relative displacement infer the difference between their respective parallaxes $(\pi_1 - \pi_2)$. When the stars are very near to each other, as seen in the sky, we refer to them as forming an optical double star. When the more distant component of such a system is so far away that its parallax is almost zero, the result of the differential measurements gives us π_1.

This method very largely eliminates the systematic errors of the first method (because these errors would be nearly the same for two neighboring stars). It appeared so promising to W. Herschel, near the end of the eighteenth century, that he undertook the systematic observation of a large number of visual double stars. The result of his work was not the meas-

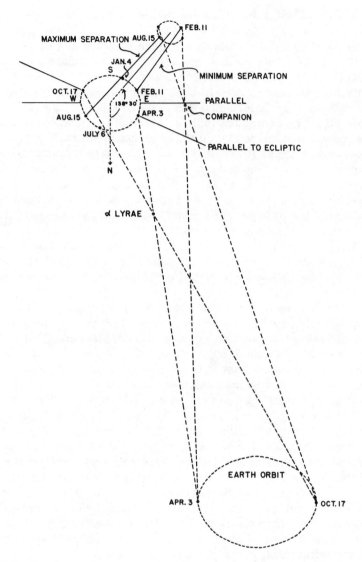

Fig. 6. The method of differential measurements, shown schematically for the star α Lyrae. The two parallactic circles must be imagined projected against the celestial sphere. In reality the distance of the companion is very much greater than that of α Lyrae; and the size of the earth's orbit has been greatly exaggerated

urement of a stellar parallax, but the discovery that most visual double stars are physical systems: the distances of the components from the sun are almost identical.

There are, of course, a number of real optical systems, but their discovery had to await the accumulation of a large amount of observational work. Both α Lyrae and 61 Cygni are such systems (the latter, in addition, is a well-known physical pair). In the former, a faint star at a distance of 43" fails to show the proper motion of α Lyrae: the latter moves with respect to the former in a straight line and with a uniform velocity. The circumstances of α Lyrae are demonstrated schematically in Figure 6. We must think of the two parallactic ellipses as being projected against the background of the celestial sphere. We notice that the distances are at maximum on August 15 of each year and at minimum on February 11. The real distance of the companion is, however, much greater than is shown in the figure, relative to the distance of α Lyrae. There is also a periodic change in the angle of the line joining the two components. But this effect is small because in reality the angular distance between the components is almost two hundred times greater than the major axis of the larger parallactic ellipse. In addition, the measurements of position angle are affected by systematic errors of measurement that depend upon orientation of the star in the sky during the time of observation. The errors in distance are known to be less serious.

Another, accidental, discovery--that of aberration--was made by James Bradley who attempted to find the parallax of γ Draconis from absolute observations of this star's celestial coordinates. The effects of aberration and parallax both cause the stars to describe small ellipses in the sky. This is shown in Figures 7 and 8.

Let us consider the star in the plane of the earth's orbit (the ecliptic). When the earth is in E_1 the star is seen in S_1 on the celestial sphere; when the earth is in E_2 the star is in S_2, and so forth. During the year the star describes a straight line from S_1 to S_2 to S_3 and back to S_4 and S_1. The total amplitude of this oscillation is less than 1 second of arc, except in α Centauri for which it is 1 1/2 seconds of arc.

But there is a much larger annual oscillation, also in a

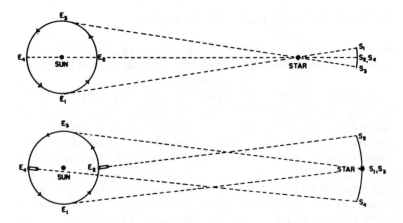

Fig. 7. *Top:* parallactic displacement of a star in the plane of the ecliptic; *bottom:* aberrational displacement of a star in the plane of the ecliptic

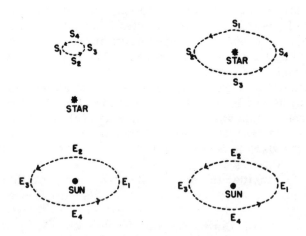

Fig. 8. *Left:* parallactic circle (seen obliquely) of a star at the pole of the ecliptic; *right:* aberrational circle (seen obliquely) of a star at the pole of the ecliptic

straight line, which is caused by the phenomenon of the aberration of light. When the earth is in E_2 it is necessary to incline the telescope in the forward direction of the earth's motion by 20 seconds of arc. This is the angle formed by the hypotenuse of a parallelogram whose sides are the velocity of light (300, 000 km/sec) and the orbital velocity of the earth (30 km/sec) respectively.

The parallactic and aberrational oscillations need never be confused because the former requires that the star be at the greatest distance from its mean position when the earth is in E_1 and E_2 (quadratures); the aberrational oscillation, on the other hand, reaches maximum displacements when the earth is in E_2 (opposition to the sun) and in E_4 (conjunction with the sun). The entire range of the aberrational oscillation is more than 40 seconds of arc.

For a star near the pole of the ecliptic, both the parallactic and aberrational oscillations become small circles on the celestial sphere, the former having a radius of less than about 1/2 second of arc, the latter one of about 20 seconds of arc. However, here it is also easy to distinguish between the two because of the fact that when the earth is in E_1 the star occupies a different position in the two celestial circles: the location of S_1 in the aberrational circle follows that of S_1 in the parallactic circle by one-fourth of a year.

History of Parallax Measurements

The Copernican theory of the solar system encountered a serious obstacle in Aristotle's comment that a moving earth would cause apparent displacements among the stars, the angle of vision under which a star is seen being different at different points along the earth's orbit. Since no such displacements had ever been detected, Aristotle rejected the theory of a moving earth, and all of his later proponents, including Tycho Brahe, believed that the absence of a measurable parallax displacement violated the Copernican concept. The early astronomers were not prepared to place the stars at those distances that were needed to make the parallactic angle too small to be detected.

Despite this objection, the Copernican theory gradually

gained ground, and by the middle of the seventeenth century
it was so well established that astronomers began to search
systematically for the displacements predicted by the theory.
One of the first of these attempts was made by Robert Hooke
in London, during the year 1669. With a telescope mounted
vertically in his house, and with the objective firmly attached
to the roof, he attempted to determine as accurately as possible the departure from the vertical plumb line of the direction
to a star that passes near the zenith of London. This small
angle would be expected to vary slightly through the seasons
(method I of section II), and Hooke's intention was to find an
annual periodicity of the kind required by the theory of parallax. The star chosen by Hooke was γ Draconis--the same star
which later led James Bradley to the discovery of the phenomenon of the aberration of light. Unfortunately Hooke's
measurements were interrupted soon after their inception
through the accidental destruction of his telescope objective.

In the years 1701 to 1704 Ole Roemer attempted to measure
the parallaxes of stars at an observatory in Copenhagen. He
chose the bright stars Sirius and Vega, which are located in
opposite points of the sky as seen from the earth. Figure 5
illustrates the method which he used. Roemer thought that he
could actually detect a small change in the angle between the
two stars, and Horrebow, who in 1727 recomputed the data
of Roemer, announced in a paper entitled "Copernicus Tri-
umphans" that the results had finally provided the lacking
confirmation of the Copernican theory. However, much later
C. A. F. Peters demonstrated conclusively that the small
angle measured by Roemer and computed by Horrebow was
in reality caused by an error in Roemer's clock, which depended upon the daily change of the temperature in the observatory.

A few years after the observations of Roemer, an amateur
astronomer named Molyneux started a series of observations
of the star γ Draconis at his private observatory near London.
With the help of his friend, James Bradley of Oxford, he established in 1725 to 1726 that the position of γ Draconis actually underwent displacements that mirrored on the celestial
sphere the orbital motion of the earth around the sun. En-

couraged by these results, James Bradley undertook the observation of another star, 35 Camelopardalis, and soon found a conspicuous annual motion in the form of an ellipse. However, both stars failed to conform with the demands of the theory of stellar parallax, as was shown in Figures 7 and 8.

In the case of γ Draconis, Bradley expected to find a maximum angle during the month of June and a minimum angle in December. During the months of March and September, the departure from the plumb was expected to have an average value. The observations yielded the opposite. The angle was greatest in September and smallest in March and passed through the average value in June and again in December.

This strange behavior led Bradley, in 1728, to the correct explanation in terms of the finite velocity of light, which results in a small shift of the telescope in the direction toward which the earth is moving. This shift is the angle of aberration and has nothing to do with the angle of parallax. One further difference should be noted: the parallax ellipses on the sky depend upon the distances of the stars, and a nearby star would have a large parallactic ellipse, while a distant star would have a small ellipse. The aberrational ellipses are independent of the distances of the stars so that, for example, the components of an optical binary whose distances from the sun differ by a factor of ten would have identical aberrational ellipses but would have parallactic ellipses whose dimensions are in the ratio of ten to one.

Bradley's observations were accurate enough to show that the aberrational ellipse of γ Draconis was not complicated by a parallactic ellipse of more than 1" in amplitude. This result placed the average naked-eye star at a distance of at least two hundred thousand astronomical units (two hundred thousand times the distance of the sun).

There the matter rested until the beginning of the nineteenth century. In 1805, however, Piazzi at Palermo published parallaxes between 2" and 10" for α Tauri, Sirius, Procyon, and Vega. Although he mistrusted some of these results, he regarded as probable the value of 4" for Sirius. In the case of α Lyrae his parallax of 2" was further exceeded by Callandrelli's measurement in Rome, 4".4. Brinkley in Dublin found for α Lyrae 1".1, and for α Aquilae, 2".75. These large values

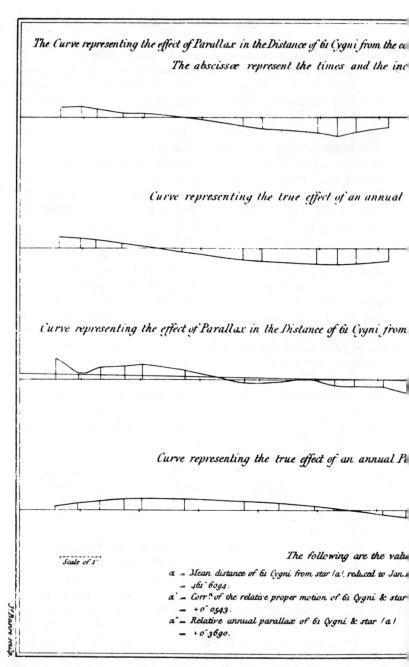

The Curve representing the effect of Parallax in the Distance of 61 Cygni from the c

The abscissæ represent the times and the inc

Curve representing the true effect of an annual

Curve representing the effect of Parallax in the Distance of 61 Cygni from

Curve representing the true effect of an annual P

Scale of 1"

J. Basire sculp.

The following are the valu

a = Mean distance of 61 Cygni from star (a), reduced to Jan. 1
  = 461˙6094 .
a′ = Corrⁿ of the relative proper motion of 61 Cygni & star
  = +0˙0543 .
a″ = Relative annual parallax of 61 Cygni & star (a)
  = +0˙3690 .

Fig. 9. Main's discussion of Bessel's observations of 61 Cygni

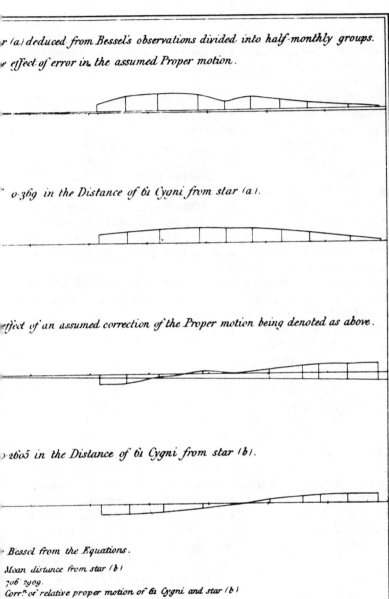

r (a) deduced from Bessel's observations divided into half-monthly groups.

e effect of error in the assumed Proper motion.

r 0·369 in the Distance of 61 Cygni from star (a).

effect of an assumed correction of the Proper motion being denoted as above.

0·2605 in the Distance of 61 Cygni from star (b).

Bessel from the Equations.

Mean distance from star (b)
706″·2909.
Corr^n of relative proper motion of 61 Cygni and star (b)
+0″·2426.
Relative annual parallax of 61 Cygni and star (b)
+0″·2606.

were criticized by Pond, at Greenwich, whose measurements indicated only that the parallaxes of most of these stars are much smaller than 1".

Bessel entered the picture in about 1812, with a rediscussion of Bradley's observations which gave for the sums of the parallaxes of pairs of stars the values:

Sirius and Vega;   0".044 ± 0".243
Procyon and Altair;   0".931 ± 0".208

In all cases the resulting parallaxes were less than 1", but the probable errors were still too large.

At the same time W. Struve obtained from his own observations of seventeen pairs of stars an average value of the parallax[21] of a single star $\pi = 0".052 \pm 0".017$. This agreed with Bessel's and Pond's results, and disagreed with those of Piazzi, Callandrelli, and Brinkley.

Even more convincing was W. Struve's discussion of the dynamical parallaxes in his 1837 volume, *Stellarum Duplicium et Multiplicium Mensurae Micrometricae*. From the application of Kepler's third law he found:

for α Virginis	$\pi M^{1/3} =$	0".185
for Castor	"	0.200
for σ Coronae Borealis	"	0.085
for ξ Ursae Majoris	"	0.147
for 70 Ophiuchi	"	0.236
for ζ Herculis	"	0.172
for 61 Cygni	"	0.254

M designates the total mass of the binary and is unknown in a specific case but may be estimated from statistical considerations for a group of stars. We *now* know that it is quite reasonable to adopt, on the average, $M = 3$ solar masses. The cube root of 3 is 1.44. Hence, the actual values of parallaxes should be about 50 per cent smaller than those listed. The best modern trigonometric values, in the order given, are 0".101, 0".072, 0".047, 0".127, 0".188, 0".110, and 0".292. The agreement is quite remarkable. But it was not a solution of the problem, since in 1837 the masses of the stars were unknown.

The actual, direct, measurements of 61 Cygni by Bessel[22], and those of α Centauri by Henderson[23] were fully discussed by R. Main in 1840.[24] I reproduce here only Main's illustration of the plotted results of the observations, with time as the abscissa and departure from mean position of 61 Cygni as the ordinate (Figure 9). These curves agree so well with the theoretical curves (designated "true" in the figure) that there remains no doubt of the reality of this parallax. Main also discussed Struve's complete series of observations of α Lyrae and his plots are in Figure 10. Although the precision is less satisfactory, Main concludes that the parallax is real. But this refers to the observations from August 16, 1836, until August 18, 1838.[25] Struve's first announcement in the *Mensurae Micrometricae* of 1837 referred only to observations between November 3, 1835, and December 31, 1836. The question which Main has not answered, but which arouses our interest is the following: Was there already in this first series a reliable indication of a measurable parallax? This is the crucial question discussed in the recent article by Deitch. Main concludes from his own recomputation of the original observations that the parallax of α Lyrae was indeed shown by these early observations, and that, moreover, the entire material from 1835 to 1838, including the distances and also the position angles (which were omitted by Struve), yields as the final result:

$$\pi = 0\rlap{.}{''}113 \pm 0\rlap{.}{''}018,$$

a value that is close to the best modern determination.

Struve's own appraisal was a more modest one. The following is J. Jackson's translation of the Latin text in the *Mensurae Micrometricae:*

As for the parallax, we nave found $\pi = + 0\rlap{.}{''}125$ with a probable error of $0\rlap{.}{''}055$. We can therefore conclude that the parallax is very small, as it probably lies between $0\rlap{.}{''}07$ and $0\rlap{.}{''}18$. But, indeed, we cannot give it yet absolutely. However, a continuation of the observations, if all are made at the epoch of maximum displacement, will draw still closer the limits of uncertainty; whence it appears indubitable that stellar parallaxes which are not less than $0\rlap{.}{''}1$ can be detected by our instrument by the method we have given.[26]

In a report to the president of the Russian Academy of Sciences, Count d'Ouvaroff, Struve wrote:

My observations give the distances and position angles of these two stars on 17 nights--which resulted in 34 equations, from which the parallax was found by the method of least squares, as $0\rlap{.}{''}125$ with a probable error of $\pm 0\rlap{.}{''}055$. This result is important because it dem-

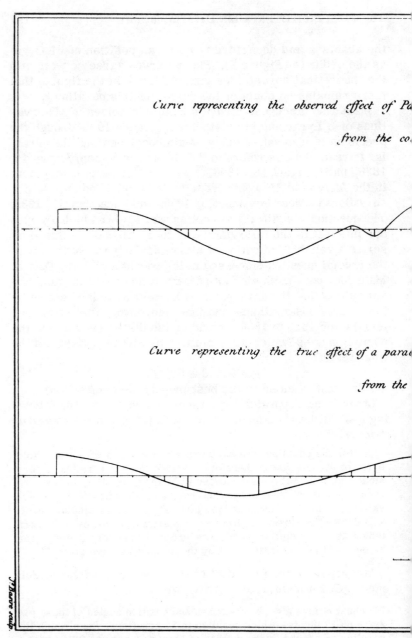

Curve representing the observed effect of Pa

from the co

Curve representing the true effect of a para

from the

J. Basire sculp.

Fig. 10. Main's discussion of Struve's observations of α Lyrae

in the measures of the Distance of a Lyræ

n star.

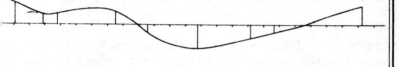

261 in the measures of Distance of a Lyræ

son star.

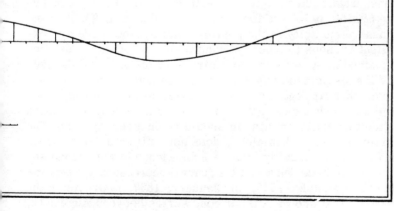

onstrates that the parallax can only be a small fraction of a second, and that consequently the results of Piazzi, Callandrelli and Brinkley were incorrect, since they had found parallaxes of the order of several seconds. On the other hand, my observations give a value that is sufficiently definite and, though small, is considerably larger than its inherent uncertainty.[27]

Bessel's opinion was given in his article on 61 Cygni:

Upon this beginning Struve bases the hope that he may be able to establish the annual parallax of α Lyrae within very narrow limits; a hope which we must regard as justified. Even from this beginning it is obvious that the observations decidedly lean in support of Pond, and consequently against Brinkley's much larger value of the annual parallax of the same star.[28]

And in a letter to Olbers dated October 18, 1837, he writes: "Struve has thus overtaken me in that he has made an attempt which, though not yet a complete success, nevertheless apparently gives strong hope that success will be achieved."[29]

This estimate agrees essentially with W. Struve's own, and with that recorded by his son " . . . that with this instrument (the new 9-inch refractor at Dorpat) . . . he finally succeeded in carrying out the first determination of the distance of a fixed star (α Lyrae) from our solar system, or at least to bracket it within narrow limits."[30]

In order to form an independent opinion I present in Figure 12 separately the observations of the two series of observations, uncorrected for linear relative proper motion of α Lyrae (Main's plots are the corrected distances). There is no annual periodicity in the position angles, and there are excellent reasons for disregarding them--as has already been explained in Struve's two memoirs. The distances of the second longer series show the distinct annual trend: there is a maximum near September 1 (the theoretical maximum as shown in Figure 6 should be on August 15) and a minimum near March 1 (the theoretical minimum should be on February 11). The earlier series, taken alone, does not suffice to demonstrate this annual periodicity. But with knowledge of the second series, there is no doubt that the general decrease of the separation between July, 1836, and January, 1837, is precisely the same as that shown in the second series between August 15, 1837, and March 15, 1838. It is thus certain that there was

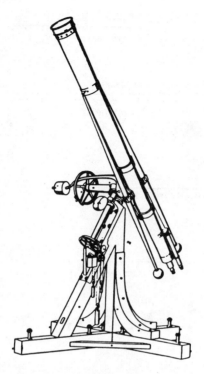

Fig. 11. Struve's Dorpat refractor--the largest refracting tele-scope in the world, when produced by Fraunhofer in 1825. Its aperture is 9 inches. (Plate from *Astronomische Nachrichten*, III [1825], 18)

a real indication of the parallax in the earlier series, but it is also certain that this indication, standing alone, would not have justified a more optimistic view than the one expressed by Bessel.

Conclusions

1. Before 1837 there were two diametrically opposite trends: Piazzi, Callandrelli, Brinkley, and others believed that they had found stellar parallaxes of between 1" and 10". Thus, Callandrelli's parallax of α Lyrae was 4".4. At the same time

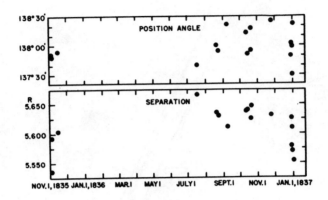

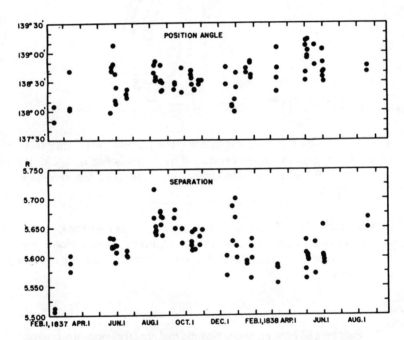

Fig. 12. Plot of Struve's observations of α Lyrae, with respect to its optical companion

Fig. 13. Portrait of F. W. Bessel. (Taken from *Abhandlungen von F. W. Bessel*, ed. R. Engelmann [Leipzig, 1875], Vol. I)

W. Struve and Bessel insisted that no star in the northern sky had a parallax in excess of a few tenths of a second.

2. In 1837 W. Struve published the results of two years of observations of α Lyrae (1835 to 1836). The parallax of 0".125 was regarded by the author as evidence that the real distance of this star was now known within narrow limits.

3. To contemporary astronomers this evidence was not fully convincing, in view of the many disappointments that had occurred in connection with previous announcements. But it led Bessel to express the *Hoffnung* that parallaxes could, in fact, be measured, and it inspired him to pursue his heliometer measurements of 61 Cygni--a star for which he had previously found a negative parallax.

4. Bessel's observations of 61 Cygni were started in 1837 and continued until 1838. The result was published in 1838 in the *Astronomische Nachrichten,* and it immediately dispelled all lingering doubts about the reality of stellar parallaxes.

5. Henderson's observations of α Centauri were made from April, 1832, to May, 1833, and therefore preceded those of Struve and Bessel. He acknowledged that Bessel had called his attention to this star. His instrument was a mural circle-- of the same construction as other mural circles which had, time after time, failed to give reliable parallaxes. Henderson was therefore not prepared to discuss and publish his results until after the positive parallaxes of α Lyrae and 61 Cygni had become known.

6. In 1840 R. Main subjected to a careful scrutiny the investigations of Bessel, Henderson, and Struve. By this time Struve had (in 1839) published the extension of his observations of α Lyrae, up to and including the year 1838. Main concluded that Bessel's results left no doubt of the reality of the parallax of 61 Cygni. He accorded the same recognition to Struve's complete series on α Lyrae, which followed in time the publication of Bessel's paper, and also to the work of Henderson on α Centauri.

7. We have the advantage of knowing that the modern value of the parallax of α Lyrae almost exactly coincides with Struve's 1837 value. An inspection of the plotted observations leaves now no doubt that the measured distances of α Lyrae and its optical companion in 1835 and 1836 show a maximum near August and a minimum near February. Thus, the existence of a stellar parallax of the order of $0''.1$ or $0''.2$ is indicated by these observations. But it is also understandable that to Main and Bessel these observations proved little more than a *Hoffnung* of success.

8. Struve's later observations which gave $\pi = 0''.261$ actually disagree with the presently accepted value to such an extent that our present appraisal must be more conservative even than Main's. These observations undoubtedly contain within them an indication of the reality of a parallax of $0''.1$ or $0''.2$, because they do show the annual oscillation demanded by the theory, but they do not carry the same conviction as Bessel's observations of 1838. The latter were not only confirmed,

almost identically, by his own later series of observations, but they agree closely with the best modern value.

9. Struve himself explained the discordance between his 1837 and 1839 results as having been caused by the action of the probable errors of the observations. No one has since succeeded in giving a better explanation. Deitch accepts it, together with the comment that small systematic errors existed in the distances, as well as in the position angles. One is tempted to believe that Struve's preoccupation with the construction and organization of the Pulkovo Observatory may have prevented him from devoting his full attention to the problem of α Lyrae. It is quite unlikely that previous knowledge of his 1837 results had induced Struve to exaggerate the parallactic oscillation in the process of making the later measurements: the immediate quantities read off at the telescope are complicated functions of several instrumental factors and also of proper motion, refraction, aberration, and so forth.

10. The best possible present appraisal is the following: Both Bessel and Struve had established before 1837 that the large parallaxes found by other astronomers were spurious. Both persisted in their efforts to establish a measurable parallax. Both were aided by two excellent Fraunhofer instruments, the Königsberg heliometer and the Dorpat refractor. To Struve belongs the credit for having found the first *trace of a real parallax* of the order of 0".1 or 0".2 in α Lyrae and of having thereby inspired Bessel[31] to continue his work on 61 Cygni. Struve also in 1837 developed the method of dynamical (or "hypothetical") parallaxes, and had given a list of double stars with their distances inferred from their orbital motions and their assumed masses. Bessel reached a precision unsurpassed in any previous work, in his measurements of 61 Cygni; and to him, without doubt, goes the credit for having produced in 1838 the *first* fully convincing determination of a star's distance. Henderson's success is impressive because he was not aided by a Fraunhofer telescope, because accidentally he hit upon the one star that even now is regarded as the sun's closest neighbor, and because the traces of its large real parallax were present in his observations before Struve and Bessel had even commenced their work.

11. The parallax of 61 Cygni is nearly three times as large

as that of α Lyrae. Was Bessel luckier or more clever than Struve in the choice of stars? Of course, α Centauri was not observable from a northern station. In reality, both northern stars were on the lists of both astronomers, and Bessel had previously observed α Lyrae without having obtained a measurable parallax. With our present knowledge of galactic structure we know that the choice of 61 Cygni, with its large proper motion and measurable orbital motion, was better than the choice of α Lyrae, which is distinguished by great apparent luminosity and a smaller proper motion. But 115 years ago the very basis of our present knowledge--the stellar parallaxes--was lacking; and the choice was a question of accident. [32]

NOTES

Egon Brunswik: Ontogenetic and Other Developmental Parallels to the History of Science

1. G. Stanley Hall, *Adolescence* (New York: D. Appleton and Co., 1904).
2. H. Werner, *Comparative Psychology of Mental Development* (rev. ed.; Chicago: Follett Publishing Co., 1948).
3. J. Piaget, *The Child's Conception of the World* (New York: Harcourt, Brace and Co., 1929), and J. Piaget, *The Child's Conception of Physical Causality* (New York: Harcourt, Brace and Co., 1930).
4. S. Bernfeld, "Zur Revision der Bioanalyse," *Imago*, XXIII (1937), 197-236.
5. Werner, *Comparative Psychology*.
6. Harry Harlow, "Thinking," *Theoretical Foundations of Psychology*, ed. H. Helson (New York: D. Van Nostrand Co., 1951), pp. 452-505.
7. F. Sander, "Structure, Totality of Experience, and Gestalt," *Psychologies of 1930*, ed. C. Murchison (Worcester, Mass.: Clark University Press, 1930), chap. x.
8. E. Brunswik, "Experimente über Kritik: Ein Beitrag zur Entwicklungs-Pyschologie des Denkens," *Bericht, XII Kongress für Psychologie*, ed. G. Kafka (Jena: Fischer, 1932).
9. Else Frenkel-Brunswik, "Psychoanalysis and the Unity of Science," *Proceedings of the American Academy of Arts and Sciences*, LXXX, No. 4 (1954).
10. J. Jastrow, ed., *The Story of Human Error* (New York: Appleton-Century, Inc., 1936).
11. G. Murphy, *Historical Introduction to Modern Psychology* (rev. ed.; New York: Harcourt, Brace and Co., 1949).

208   Notes

12. See Brunswik, "Experimente über Kritik."
13. W. Köhler, *Gestalt Psychology* (New York and London: H. Liver-
ight, 1929).
14. P. Frank, "Foundations of Physics," *International Encyclopedia
of Unified Science* (Chicago: University of Chicago Press), Vol. I, No.
7 (1946).
15. H. Feigl, "Logical Empiricism," *Readings in Philosophical Analysis*, ed. H. Feigl and W. Sellars (New York: Appleton-Century-Croft,
Inc., 1949).
16. E. Brunswik, "The Conceptual Framework of Psychology," *International Encyclopedia of Unified Science* (Chicago: University of Chicago Press), Vol. I, No. 10 (1952).
17. Piaget, *Child's Conception of the World;* Piaget, *Child's Conception of Causality.*
18. M. Wertheimer, *Drei Abhandlungen zur Gestalttheorie* (Erlangen:
Philosophische Akademie, 1925).
19. J. Piaget and B. Inhelder, *Le développement des quantités chez
l'enfant: Conservation et atomisme* (Neuchatel: Delachaux, 1941); J.
Piaget and A. Szeminska, *Le genèse du nombre chez l'enfant* (Neuchatel:
Delachaux, 1941).
20. E. Brunswik, *Wahrnehmung und Gegenstandswelt* (Leipzig and
Vienna: Deuticke, 1934).
21. Werner, *Comparative Psychology.*
22. For a summary of some of the quantitative results see Charlotte
Bühler, *Kindheit und Jugend* (3rd ed.; Leipzig: Hirzel, 1931).
23. P. Frank, *Modern Science and its Philosophy* (Cambridge, Mass.:
Harvard University Press, 1949).
24. Brunswik, "Conceptual Framework of Psychology."
25. A. Willwoll, *Begriffsbildung: Eine psychologische Untersuchung*
(Leipzig: Hirzel, 1926).
26. Frank, *Modern Science.*
27. S. C. Pepper, *World Hypotheses: A Study in Evidence* (Berkeley:
University of California Press, 1948).
28. K. Lewin, "The Conflict Between Aristotelian and Galileian Modes
of Thought in Contemporary Psychology," *Journal of General Psychology*,
V (1931), 141-77; also in K. Lewin, *A Dynamic Theory of Personality*
(New York: McGraw-Hill Book Co., 1935).
29. As Clyde Kluckhohn informs me (private communication), similar views are common among the Navajos.
30. Köhler, *Gestalt Psychology.*
31. Sander, "Structure, Totality of Experience, and Gestalt."
32. Brunswik, "Conceptual Framework of Psychology."
33. *Ibid.,* pp. 50 ff.
34. G. W. Allport, "The Psychologist's Frame of Reference," *Psychological Bulletin*, XXXVII (1940), 1-28; J. S. Bruner and G. W. Allport, "Fifty Years of Change in American Psychology," *ibid.,* pp. 757-
76.

Stillman Drake: J. B. Stallo and the Critique of Classical Physics

1. Thomas J. McCormack, "John Bernard Stallo," *Open Court*, XIV, No. 5 (May, 1900), 276, 283.

2. J. B. Stallo, *The Concepts and Theories of Modern Physics* (New York: D. Appleton and Co. 1882), p. 7.

3. H. A. Rattermann, *Johann Bernhard Stallo, deutsch-amerikanischer Philosoph, Jurist und Staatsmann: Denkrede gehalten im Deutschen Literarischen Klub von Cincinnati am 6 November 1901* (Cincinnati: Verlag des Verfassers, 1902). This pamphlet is the principal source of biographical material concerning Stallo, and all subsequent quotations ascribed to Rattermann are translated from it.

4. J. B. Stallo, *Die Begriffen und Theorien der modernen Physik. Nach der 3. Auflage des englischen Originals übersetzt und herausgegeben von Dr. Hans Kleinpeter. Mit einem Vorwort von Ernst Mach* (Leipzig: J. A. Barth, 1901).

5. Stallo, *Begriffen*, p. iv (my translation).

6. Bertrand A. W. Russell, *An Essay on the Foundations of Geometry* (Cambridge, Eng.: Cambridge University Press, 1897), p. 109 footnote.

7. Stallo, *Begriffen*, p. iii (my translation).

8. *Ibid.*, pp. ix-xiii (my translation).

9. J. B. Stallo, *Reden, Abhandlungen und Briefe* (New York: E. Steiger and Co., 1893), pp. 77 ff.

10. Ernst Mach, *The Science of Mechanics*, trans. T. J. McCormack (5th Eng. ed.; La Salle, Ill.: The Open Court Publishing Co., 1942), p. 341. The original German edition, published in 1883, was apparently not known to Stallo when he answered his critics in the following year.

11. J. B. Stallo, "The Primary Concepts of Modern Physical Science," *Popular Science Monthly*, IV (December, 1873), 230. Cf. Stallo, *Concepts* (1882), pp. 200-1.

12. Stallo, *Concepts* (1891), pp. iii-iv.

13. Stallo, *Concepts* (1882), p. 88.

14. *Ibid.*, p. 108.

15. *Ibid.*, p. 107.

16. H. Kleinpeter, "J. B. Stallo als Erkenntniskritiker," *Vierteljahrsschrift für wissenschaftlichen Philosophie*, XXV, No. 4 (1901), 401-40. Stallo did not, however, die entirely without recognition. Two books at least were dedicated to him: Thomas Sterry Hunt, *A New Basis for Chemistry* (New York: Scientific Publishing Co., 1887), and E. Mach, *Die Prinzipien der Wärmelehre historisch-kritisch entwickelt* (2nd. ed.; Leipzig: J. A. Barth, 1900). At least one eminent American physicist, Samuel Pierpont Langley, was impressed by Stallo's work; see H. Adams, *The Education of Henry Adams* (Modern Library ed.; New York: Random House, 1931, p. 377. While the present paper was in press, I was informed that Professor P. W. Bridgman is preparing a new edition of the *Concepts* to be published at Harvard.

17. Stallo, *Concepts* (1882), p. 11.
18. W. J. Youmans, "Sketch of John B. Stallo," *Popular Science Monthly,* XXXIV (February, 1889), 553.
19. Kleinpeter, "J. B. Stallo als Erkenntniskritiker," p. 403-4.
20. *Popular Science Monthly,* III (October, 1873), 771-72.

E. O. Essig: Charles Fuller Baker, American Entomologist, Botanist, Teacher, 1872-1927

1. *Colorado Agricultural Experiment Station,* Bulletin No. 31, Technical Series 1, pp. 1-137. Baker's principal earlier papers concerned Michigan Araneae *(Entomological News,* V [1894], 163-64), Colorado Coleoptera *(ibid.,* VI [1895], 173-74), and Colorado Diptera *(ibid.,* pp. 27-29). He was a prolific writer and published in all approximately 183 papers most of which are listed by Basilio Hernandez in "A Bibliography of the Published Writings of Dean Charles Fuller Baker from 1894 to 1927," *Lingnan Science Journal* (Canton, China), V, no. 3 (1927), 271-79, nos. 1-162, plus a mimeographed supplement which is attached, nos. 163-72 (1897), 1910, and 1928).
2. According to Dr. Paul W. Oman, Bureau of Entomology and Plant Quarantine, United States Department of Agriculture, the proper scientific designation now is *Circulifer tenellus* (Baker). It is believed to have been introduced into the United States from the Mediterranean region and now occurs throughout the Rocky Mountain region and westward to the Pacific Ocean.
3. William E. Hoffmann, "Charles Fuller Baker," *The Lingnaam Agricultural Review,* IV, No. 2 (October, 1927), 197-202.
4. *Invertebrata Pacifica* (Claremont, California, and Havana, Cuba), ed. with contributions by C. F. Baker, I, 12 parts (1903-7), 1-197.
5. "A Revision of American Siphonaptera, or Fleas, Together with a Complete List and Bibliography of the Group," *Proceedings of the United States National Museum,* XXVII (1904), 365-469.
6. "The Classification of the American Siphonaptera," *ibid.,* XXIX (1905), 121-61.
7. Franklin Sumner Earle, "A Letter from F. S. Earle," in Memoriam Issue, *The Philippine Agriculturist,* XVI (Special Number, 1928), 44-45.
8. Hoffman, "Charles Fuller Baker," pp. 197-202.
9. Charles W. Hamilton, "The Funeral Addresses," in Memoriam Issue, *The Philippine Agriculturist,* XVI (Special Number, 1928), 15-17.
10. Hoffmann, "Charles Fuller Baker," pp. 197-202.
11. Edwin Bingham Copeland, "Charles Fuller Baker, A Tribute," in Memoriam Issue, *The Philippine Agriculturist,* XVI (Special Number, 1928), 8-10.
12. Colin G. Welles, "Charles Fuller Baker--a Sketch," *Science,* LXVI, No. 1706 (September 9, 1927), 229-30.
13. Some of the specialists who were supplied with entomological ma-

terial by Baker: Coleoptera: Hans Gebien, R. Kleine, and A. Zimmermann, Germany; A. Boucomont, Ed. Fleutiaux, A. Grouvelle, and M. Pic, France; Jan Obenberger, Czechoslovakia; H. Krekich-Strassoldo, Austria; Chr. Aurivillius, Sweden; H. E. Andrewes and Guy A. K. Marshall, England; Edward A. Chapin, United States. Orthoptera: Achile Griffini, Italy; H. H. Karny, Dutch East Indies; A. N. Caudell, United States. Homoptera: Frederick Muir and D. L. Crawford, Hawaii; L. Melichar, Moravia; W. D. Funkhouser and T. D. A. Cockerell, United States. Hemiptera: W. L. McAtee and J. R. Malloch, United States. Diptera: P. Sack, Germany; W. S. Patton, Scotland; M. Bezzi, Italy; J. R. Malloch and G. F. Ferris, United States. Hymenoptera: E. A. Elliott, England; H. L. Viereck, Canada; T. D. A. Cockerell, United States. Baker worked chiefly in Homoptera on the Jassoidea, Fulgoridae, and Cercopidae, and in the Hymenoptera on the parasitic Braconidae, during his stay in the Philippines.

14. Welles, "Charles Fuller Baker--a Sketch," pp. 229-30.

15. Dr. Robert Larimore Pendleton was born in Minneapolis, Minnesota, June 25, 1890. He graduated from the Department of Soils, College of Agriculture, University of California, with a B.S. in 1914 and a Ph.D. in 1917. From 1918 to 1942 he was the leading soil technologist in India, the Philippine Islands, Siam, China, and other countries of the Far East. He was intimately associated with Baker when he was professor of soil technology and the latter was dean of the College of Agriculture at the University of the Philippines. He and Baker were the last of the Americans to be replaced by Filipinos. Upon leaving the Far East he became principal soil technologist in the office of Foreign Agricultural Relations, United States Department of Agriculture, 1935-42. In 1942 he accepted the position of professor of tropical soils and agriculture in the Isaiah Bowman School of Geography at Johns Hopkins University, a position he now holds. (American Men of Science, ed. by Jacques Cattell, 8th ed. [Lancaster, Penna.: The Science Press, 1949], p. 1921.)

16. As nearly as can be ascertained, and according to Baker's own estimate, his insect collection contained about 250,000 specimens in all orders. Cushman, who packed the collection, estimated it had more nearly 300,000 specimens, with perhaps 50,000 still in the hands of some 115 collaborating specialists mostly in Europe. It was kept in 1,417 standard 13 x 9 inch Schmitt boxes distributed by orders as follows: Orthoptera 96, Dermaptera 5, Plecoptera 1, Corrodentia 1, Odonata 4, Homoptera 208, Heteroptera 115, Neuroptera 6, Trichoptera 6, Lepidoptera 125, Coleoptera 607, Hymenoptera 207, Diptera 40. The card catalogue, previously estimated at 35,000, actually consisted of about 100,000 cards.

Frederick O. Koenig: On the History of Science and the Second Law of Thermodynamics

1. J. R. Oppenheimer, Science and the Common Understanding (London: Oxford University Press, 1954), p. 95.

2. The only biography of Sadi Carnot known to me at the time of writing the above (1954), that by his brother, Hippolyte (H. Carnot, "Notice biographique sur Sadi Carnot" in *Réflexions sur la puissance motrice du feu et sur les machines propres a développer cette puissance* by N. L. S. Carnot [2nd ed.; Paris: Gauthier-Villars, 1878]), says nothing on the subject. Since then I have learned from Dr. T. S. Kuhn of the complete edition of Sadi Carnot's posthumous notes by E. Picard in *Sadi Carnot, biographie et manuscript* (Paris: Gauthier-Villars, 1927). This might shed some light on the matter.

3. (Paris: Bachelier, 1824), hereafter referred to as *Réflexions*. All of my quotations (in English) from the *Réflexions* are taken from the translation by W. F. Magie in *The Second Law of Thermodynamics* (New York and London: Harper and Bros., 1899).

4. There are some good beginnings, e.g., a recent penetrating and suggestive study by S. Lilley, *Social Aspects of the History of Science* (Paris: Peyronnet and Co., undated), which contains a discussion of the First Law of thermodynamics.

5. What may seem a needless digression is perhaps further justified by a recent biography of Lazare Carnot, by S. J. Watson, a British army officer, which makes no mention of the great man's scientific importance *(Carnot, 1752-1823* [London: The Bodley Head, 1954]). My data are taken from an excellent short work by K. Fink, a German *Gymnasiallehrer (Lazare-Nicolas-Marguerite Carnot, sein Leben und seine Werke nach den Quellen dargestellt* [Tübingen: H. Laupp, 1894]). A full-length biography has been given by Hippolyte Carnot in *Mémoires sur Carnot par son fils* (2 vols.; Paris: Pagnerre, 1861, 1863).

6. L. N. M. Carnot, *Principes fondamentaux de l'équilibre et du mouvement* (Paris: Chez Deterville, 1803).

7. L. N. M. Carnot, *Géométrie de position* (Paris: M. Duprat, 1803).

8. My data on the life of Sadi are taken from the all too brief essay by his brother Hippolyte, which is appended, along with Sadi's posthumous writings, to the second edition of the *Réflexions*. My quotations from this biography are from the translation by R. H. Thurston, *Reflections on the Motive Power of Heat* (New York: John Wiley and Sons, 1897).

9. For some observations on this trait from the psychological point of view, see the study by Anne Roe, *The Making of a Scientist* (New York: Dodd, Mead and Co., 1952).

10. N. L. S. Carnot, "Extrait de notes inédites de Sadi Carnot sur les mathématiques, la physique et autres sujets," *Réflexions* (2nd ed.). English translation is in Thurston, *Reflections*. The notes were published in their entirety only in 1927, by E. Picard in *Sadi Carnot*. [Added April, 1958.]

11. Biographical information on Clausius is scanty. My data are taken from the *Encyclopaedia Britannica*, 11th ed., and from an obituary by J. W. Gibbs in *The Collected Works of J. Willard Gibbs* (London: Longmans, Green and Co., 1928), II, 261. The latter consists mostly of an excellent summary of Clausius' scientific work.

12. R. J. E. Clausius,"Über die bewegende Kraft der Wärme und die Gesetze die sich daraus für die Wärmelehre selbst ableiten lassen," *Poggendorffs Annalen der Physik und Chemie*, LXXIX (1850), 368, 500; also in Part I of R. J. E. Clausius, *Abhandlungen über die mechanische Wärmetheorie* (Braunschweig: Fr. Viehweg and Son, Part I, 1864; Part II, 1867). English translation is in Magie, *The Second Law*.

13. B. P. E. Clapeyron, "Sur la puissance motrice de la chaleur," *Journal de l'École Polytechnique*, XIV (1834), 153. German translation is in K. Schreber, *Abhandlung über die bewegende Kraft der Wärme von E. Clapeyron* (Ostwald's Klassiker Nr. 216 [Leipzig: Akademische Verlagsgesellschaft, 1926]).

14. W. Thomson, "An Account of Carnot's Theory of the Motive Power of Heat; With Numerical Results Deduced from Regnault's Experiments on Steam," *Transactions of the Royal Society of Edinburgh*, XVI (1849), 541; W. Thomson, *Mathematical and Physical Papers by Sir William Thomson* (Cambridge, Eng.: University Press, 1882), I, 113.

15. Gibbs, *Collected Works*, II, 261.

16. Wilhelm Ostwald, *Grosse Männer* (Leipzig: Akademische Verlagsgesellschaft, 1909).

17. My data are taken from the biography by Sylvanus P. Thomson, *The Life of William Thomson, Baron Kelvin of Largs* (London: Macmillan Co., 1910). A valuable feature of this work is the complete bibliography of Lord Kelvin's writings.

18. *Ibid.*

19. W. Thomson, "On an Absolute Thermometric Scale Founded on Carnot's Theory of the Motive Power of Heat, and Calculated from Regnault's Observations," *Philosophical Magazine*, Series 3, XXXIII (1848), 313; W. Thomson, *Mathematical and Physical Papers*, I, 100.

20. *Transactions of the Royal Society of Edinburgh*, XVI (1849), 541; W. Thomson, *Mathematical and Physical Papers*, I, 113.

21. W. Thomson, "On the Dynamical Theory of Heat, with Numerical Results Deduced from Mr. Joule's Equivalent of a Thermal Unit, and M. Regnault's Observations on Steam," *Transactions of the Royal Society of Edinburgh*, XX (1853), 261; *Philosophical Magazine*, Series 4, IV (1852), 8, 105, 168, 424; W. Thomson, *Mathematical and Physical Papers*, I, 174-210.

22. W. Thomson, "On a Universal Tendency in Nature to the Dissipation of Mechanical Energy," *Proceedings of the Royal Society of Edinburgh*, (1852); *Philosophical Magazine*, Series 4, IV (1852), 304; W. Thomson, *Mathematical and Physical Papers*, I, 511.

23. W. Thomson, "On the Dynamical Theory of Heat, etc. Part VI. Thermo-electric Currents," *Transactions of the Royal Society of Edinburgh*, XXI (1857), 123; W. Thomson, *Mathematical and Physical Papers*, I, 232.

24. Ostwald, *Grosse Männer*.

25. These relations are evidently excellent grounds for taking the G-Law as *the* Second Law, as suggested in §7b.

26. F. O. Koenig, "On the Significance of the Forgotten Thermodynamic Theorems of Carnot," *Essays in Biology in Honor of Herbert M. Evans* (Berkeley and Los Angeles: University of California Press, 1943).

27. E. Mach, *Die Prinzipien der Wärmelehre historisch-kritisch entwickelt* (2nd ed.; Leipzig: J. A. Barth, 1900), p. 215.

28. C. Singer, *The Story of Living Things* (New York and London: Harper and Bros., 1931), p. 103.

29. N. L. S. Carnot, "Extrait de notes inédites," in *Réflexions* (2nd ed.); also Picard, *Sadi Carnot*.

30. H. L. Callendar, "Heat," *Encyclopaedia Britannica* (11th ed.; XIII 1910-11), 144.

31. Schreber, *Abhandlung über die bewegende Kraft 'der Wärme von E. Clapeyron,* p. 41.

32. J. N. Brönsted, *Philosophical Magazine,* VII, Series 29 (1940), 699.

33. V. K. LaMer, "Some Current Misconceptions of N. L. Sadi Carnot's Memoir and Cycle," *Science,* CIX (1949), 598; *American Journal of Physics,* XXII (1954), 20.

34. L. Brillouin, "La vie, la pensée et la physico-chimie," *Les Cahiers de la Pléiade,* No. 12 (1952).

35. Since I wrote the above (1954), LaMer has published a further paper in support of his interpretation, "Some Current Misconceptions of N. L. Sadi Carnot's Memoir and Cycle II," *American Journal of Physics,* XXIII (1955), 95, and T. S. Kuhn has published a thorough critical discussion of LaMer's papers and reached conclusions substantially in agreement with mine; see T. S. Kuhn, "Carnot's Version of Carnot's Cycle," *American Journal of Physics,* XXIII (1955), 91; Kuhn, "LaMer's Version of Carnot's Cycle," *American Journal of Physics,* XXIII (1955), 387. [Added April, 1958]

36. I have had no opportunity to examine the book itself. My information is derived from the account by K. Fink, *Lazare-Nicolas-Marguerite Carnot.* Especially relevant in the present connection are pp. 104-5 and 123 of Fink's work.

37. Mach, *Die Prinzipien der Wärmelehre,* p. 215.

38. J. R. Mayer, "Bemerkungen über die Kräfte der unbelebten Natur," *Annalen der Chemie und Pharmacie,* XLII (1842), 233; also Ostwald's Klassiker Nr. 180 (Leipzig: Akademische Verlagsgesellschaft, 1911).

39. J. P. Joule, *Philosophical Magazine,* XXIII, Series 3 (1843), 263, 347, 435; extract in W. F. Magie, *A Source Book in Physics* (New York and London: McGraw-Hill Book Co., 1935), p. 203.

40. H. L. F. Helmholtz, *Über die Erhaltung der Kraft* (Berlin: G. Reimer, 1847); also Ostwald's Klassiker Nr. 1 (Leipzig: Engelmann, 1889).

41. J. R. Mayer, *Die organische Bewegung in ihrem Zusammenhang mit dem Stoffwechsel* (Heilbronn: Drechsler, 1845); also Ostwald's Klassiker No. 180 (Leipzig: Akademische Verlagsgesellschaft, 1911).

42. J. R. Mayer, *Beiträge zur Dynamik des Himmels* (Heilbronn:

Landherr, 1848); also Ostwald's Klassiker Nr. 223 (Leipzig: Akademische Verlagsgesellschaft, 1927).

43. J. P. Joule, *Philosophical Magazine*, XXVI, Series 3 (1845), 369; *ibid.*, XXVII, Series 3 (1845), 205. Extract in Magie, *A Source Book*, p. 205.

44. J. P. Joule, *Philosophical Magazine*, XXXI, Series 3 (1847), 173; extract in Magie, *A Source Book*, p. 207.

45. J. P. Joule, *Philosophical Transactions of the Royal Society of London*, CXL (1850), 61. Extract in Magie, *A Source Book*, p. 207.

46. C. Holtzmann, *Über die Wärme und Elasticität der Gase und Dämpfe* (Mannheim, 1845); extract in *Poggendorffs Annalen der Physik und Chemie*, Ergänzungsband II (1848).

47. Had it been complete he would have anticipated Clausius. One may wonder at Helmholtz's feelings on realizing how narrowly, in his epochal memoir on the First Law, he had missed contributing also to the discovery of the Second.

48. In the *Mathematical and Physical Papers*, these tables, though mentioned in the paper of 1848, are presented only in that of 1849 which I refer to for other reasons in §§6b, 6c, 11, 14.

49. W. Thomson, "On the Dynamical Theory of Heat. Part VI."

50. W. Thomson, *Mathematical and Physical Papers*, Vol. I.

51. J. Thomson, "Theoretical Considerations of the Effect of Pressure in Lowering the Freezing Point of Water," *Transactions of the Royal Society of Edinburgh*, XVI (1849), 575; W. Thomson, *Mathematical and Physical Papers*, I, 156.

52. W. Thomson, "The Effect of Pressure in Lowering the Freezing Point of Water Experimentally Demonstrated," *Proceedings of the Royal Society of Edinburgh* (1850); *Philosophical Magazine*, Series 3, XXXVII (1850), 123; W. Thomson, *Mathematical and Physical Papers*, I, 165.

53. W. Thomson, "An Account of Carnot's Theory of Motive Power."

54. *Ibid.*

55. Clausius, "Über die bewegende Kraft der Wärme."

56. J. C. Maxwell, *Nature*, XVII (1878), 257.

57. This question is quoted by Gibbs in a preceding paragraph of the obituary, as follows: "'For more than twelve years,' said Regnault in 1853, 'I have been engaged in collecting the materials for the solution of this question: Given a certain quantity of heat, what is, theoretically, the amount of mechanical effect which can be obtained by applying the heat to evaporation, or the expansion of elastic fluids, in the various circumstances which can be realized in practice?'" (H. V. Regnault, *Comptes Rendus*, XXXVI [1853], 676).

58. Gibbs, *Collected Works*, II, 261.

59. Not to be confused with the *a* of equation (8). In discussing the work of any scientist I use his notation, and, therefore, the meaning of a given letter may change as we pass from one scientist to another in this article.

60. Clausius does not mention the fact that the law for the adiabatic change of a gas was already known, having been discovered (like Car-

not's Theorem [e. c.]) by deduction from the caloric theory, by Laplace and by Poisson, around 1823, and confirmed by the quantitative experimental data of several authors; see T. S. Kuhn, "The Caloric Theory of Adiabatic Compression," *Isis*, XLIX (1958), 132. [Added April, 1958]

61. W. Thomson, "On the Dynamical Theory of Heat."

62. This is the paper mentioned in §10d as that in which Thomson first introduced the Kelvin scale in use today, and in §15a as that in which he independently deduced the "equality" of Clausius' theorem.

63. Clausius, *Poggendorffs Annalen der Physik und Chemie*, XCIII (1854), 481; Clausius, *Abhandlungen über die mechanische Wärmetheorie*, Part I.

64. This is the paper already mentioned for other reasons in §§10d and 14.

65. R. J. E. Clausius,"Über eine veränderte Form des zweiten Hauptsatzes der mechanischen Wärmetheorie," *Poggendorffs Annalen der Physik und Chemie*, XCIII (1854), 481; Clausius, *Abhandlungen über die mechanische Wärmetheorie*, Part I.

66. R. J. E. Clausius, "Über die Anwendung des Satzes von der Äquivalenz der Verwandlungen auf die innere Arbeit," *Poggendorffs Annalen der Physik und Chemie*, CXVI (1862), 73; Clausius, *Abhandlungen über die mechanische Wärmetheorie*, Part I.

67. These quantities, which under other names became important for statistical mechanics, of which Clausius was a pioneer, are commented on by Gibbs in his obituary of Clausius *(Collected Works*, II, 261).

68. R. J. E. Clausius, "Über einen Grundsatz der mechanischen Wärmetheorie," *Poggendorffs Annalen der Physik und Chemie*, CXX (1863), 426; Clausius, *Abhandlungen über die mechanische Wärmetheorie*, Part I.

69. Clausius, *Poggendorffs Annalen der Physik und Chemie*, CXXV (1865), 353; Clausius, *Abhandlungen über die mechanische Wärmetheorie*, Part II.

70. M. Planck, *Über den zweiten Hauptsatz der mechanischen Warmetheorie: Inauguraldissertation* (Munich: Ackermann, 1879).

71. M. Planck, *Thermodynamik* (Leipzig: Veit and Co., 1897). For further remarks on Planck's work on thermodynamics, see the article by Victor F. Lenzen in the present volume.

72. C. Carathéodory, *Mathematische Annalen*, LXVII (1909), 355.

73. M. Born, "Kritische Betrachtungen zur Thermodynamik," *Physikalische Zeitschrift*, XXII (1921), 218, 249, 282.

74. G. N. Lewis and M. Randall, *Thermodynamics and the Free Energy of Chemical Substances* (New York and London: McGraw-Hill Book Co., 1923), p. 112.

Victor F. Lenzen: Planck's Philosophy of Science

1. See the article by F. O. Koenig in the present volume.

2. See the article by Stillman Drake in the present volume.

3. The addresses of Max Planck cited in the present essay have been published in *Vorträge und Erinnerungen* (5th ed. of *Wege zur physikalischen Erkenntnis*; Stuttgart: S. Hirzel, 1949). English translations of some items have been published in M. Planck, *Scientific Autobiography, and Other Papers*, trans. Frank Gaynor (New York: Philosophical Library, 1949). In 1958 there was published in three volumes, M. Planck, *Physikalische Abhandlungen und Vorträge* (Braunschweig: F. Vieweg and Son).

Robert H. Lowie: The Development of Ethnography as a Science

1. Berthold Laufer, "The Fundamental Ideas of Chinese Culture," *The Journal of Race Development,* V (1914), 160-74.

2. Berthold Laufer, "The Early History of Felt," *American Anthropologist,* XXXII (1930), 1-18.

3. Wilhelm E. Mühlmann, *Geschichte der Anthropologie* (Bonn, 1948), pp. 41-44; W. Schmidt and W. Koppers, *Völker und Kulturen* (Regensburg, 1924), p. 20; W. Vernon Kienitz, *The Indians of the Western Great Lakes 1615-1760* (Ann Arbor: University of Michigan Press, 1940), p. 333.

4. John R. Swanton, "Religious Beliefs and Medical Practices of the Creek Indians," in *42nd Annual Report 1924-25* (U. S. Bureau of American Ethnology [Washington, D. C.: Government Printing Office, 1928]), p. 520.

5. Hermann Hauff. *Alexander von Humboldts Reise in die Aequinoctial-Gegenden des neuen Continents* (Stuttgart, 1862), VI, 124, 282 f.; J. Moleschott, *Georg Forster, der Naturforscher des Volks* (Frankfurt a. M.: Meidinger Son and Co., 1854), pp. 47, 57.

6. A. L. Kroeber and Clyde Kluckhohn, "Culture: A Critical Review of Concepts and Definitions," *Papers, Peabody Museum of American Archaeology and Ethnology, Harvard University,* XLVII, No. 1 (1952), 10, 24 f., 35 f.

7. Erland Nordenskiöld, *Comparative Ethnographical Studies* (Göteborg: Erlanders boktryckeri aktiebolag), IX (1931), 101-12.

8. Charles Darwin, *Journal of Researches* (2nd ed.; London: J. Murray, 1845), chap. x.

9. Georg Schweinfurth, *The Heart of Africa* (London: S. Low, Marston, Low, and Searle, 1873), II, 60 f.; G. Schweinfurth, *Artes Africanae* (Leipzig and London: S. Low, Marston, Low, and Searle, 1875), Pls. XVI, XVIII; W. Scoresby Routledge and Katherine Routledge, *With a Prehistoric People: The Akikuyu of British East Africa* (London: E. Arnold, 1910), pp. 97-102.

10. Otis T. Mason, *Aboriginal American Basketry* (from U. S. National Museum Annual Report for 1902, Vol. II [Washington, D. C.: Government Printing Office, 1904]), pp. 171-548.

11. Lord Avebury, *The Origin of Civilization and the Primitive Condition of Man* (London: Longmans, Green and Co., 1911), pp. 409-31.

12. A. L. Kroeber, *The Nature of Culture* (Chicago: University of Chicago Press, 1952), p. 137; cf. *contra*, David Bidney, "The Concept of Value in Modern Anthropology, " *Anthropology Today* (New York: International Symposium on Anthropology, 1952), pp. 682-99.

13. Kroeber and Kluckhohn, "Culture, " p. 181.

14. Richard Thurnwald, *Die menschliche Gesellschaft in ihren ethnosoziologischen Grundlagen* (Berlin and Leipzig: W. de Gruyter and Co. , 1935), IV, 321.

15. Lewis H. Morgan, *Ancient Society* (New York: H. Holt and Co. , 1877), Part I, chap. i; Part II, chap. xiv.

16. Schmidt and Koppers, "Völker und Kulturen, pp. 210, 264 f.

17. V. Gordon Childe, "Archaeological Ages as Technological Stages" (Huxley Memorial Lecture), *Journal of the Royal Anthropological Institute of Great Britain and Ireland*, LXXIV (1944), 7-24.

18. Julian H. Steward, "Cultural Causality and Law: A Trial Formulation of the Development of Early Civilizations, " *American Anthropologist*, LI (1949), 1-27.

19. E. B. Tylor, "On a Method of Investigating the Development of Institutions, " *Journal of the Royal Anthropological Institute of Great. Britain and Ireland*, XVIII (1889), 245-69.

20. George Peter Murdock, *Social Structure* (New York: Macmillan Co. , 1949).

21. Harold E. Driver, "Statistics in Anthropology, " *American Anthropologist*, LV (1953), 42-59.

22. A. R. Radcliffe-Brown, *The Social Organization of Australian. Tribes* (Melbourne and London: Macmillan Co. , 1931); Claude Lévi-Strauss, *Les structures élémentaires de la parenté* (Paris: Presses Universitaires de France, 1949); Fred Eggan, *Social Organization of the Western Pueblos* (Chicago: University of Chicago Press, 1950).

23. Eggan, *Social Organization*, p. 8.

Leonardo Olschki: Main Topics in Marco Polo's Description of the World

1. Critical edition by Luigi Foscolo Benedetto, *Il Milione* (Florence: Leo S. Olschki, 1928). English translations and commentaries: Henry Yule, *The Book of Ser Marco Polo, the Venetian, Concerning the Kingdoms and Marvels of the East* (3rd ed. rev. by Henri Cordier; 2 vols. ; New York: Charles Scribner's Sons, 1921), with a supplementary volume by H. Cordier, *Ser Marco Polo Notes and Addenda* (London: John Murray, 1920). A. C. Moule and P. Pelliot, trans. and annot., *Marco Polo: The Description of the World* (2 vols. ; London: Routledge, 1938). Volume I is a translation *variorum* with introduction, notes, and appendix; volume II, the Latin text of the rediscovered Zelada manuscript of the Cathedral library of Toledo, Spain. A planned third volume with a com-

mentary by Paul Pelliot has not yet appeared. Marco Polo, *The Travels of Marco Polo*, trans. W. Marsden, introduction by John Masefield (Everyman's Library ed.; New York: E. P. Dutton and Co., 1932). *The Most Noble and Famous Travels of Marco Polo, Together with the Travels of Nicolò de'Conti*, edited from the Elizabethan translation of John Frampton, with introduction, notes, and appendixes, by N. M. Penzer (London: The Argonaut Press, 1929).

2. Translation of this preface by Yule, *Book of Ser Marco Polo*, II, 525.

3. See John K. Wright, *The Geographical Lore of the Time of the Crusades* (New York: American Geographical Society, 1925), and C. V. Langlois, *La connaissance de la nature et du monde au moyen âge* (Paris: Librairie Hachette, 1927), chap. v. See also L. Olschki, *Marco Polo's Precursors* (Baltimore, Md.: The Johns Hopkins Press, 1943).

4. Francesco Balducci Pegolotti, *Pratica della Mercatura*, ed. Allan Evans (Cambridge, Mass.: Mediaeval Academy of America, 1936).

5. Yule, *Book of Ser Marco Polo*, II, 236.

6. *Ibid.*, pp. 312, 424.

7. *Ibid.*, Introduction, p. 120, and for Peter of Abano see Lynn Thorndike, *A History of Magic and Experimental Science* (New York: The Macmillan Co., 1923), II, 874-947.

8. The "Star" is, of course, a constellation. There are more examples of "stella" used in this sense in medieval literature.

9. Cf. Friar John's first letter from "Mabar" (southern India) of December 30, 1292 or 1293, just when Marco Polo journeyed in the Indian Ocean on his way back from China to Venice, in the original Italian text in *Sinica Franciscana*, ed. P. Anastasius van den Wyngaert (Quaracchi-Florence: Collegio San Bonaventura, 1929), I, 341, §4.

10. Samuel E. Morison, *Admiral of the Ocean Sea: A Life of Christopher Columbus* (Boston: Little, Brown and Co., 1942), II, 282 ff., and for Vespucci's controversial astronomical data, see Frederick J. Pohl, *Amerigo Vespucci* (2nd ed.; New York: Columbia University Press, 1945), pp. 147 ff.

11. For the principal details of this event see W. W. Rockhill, *The Journey of William of Rubruck to the Eastern Parts of the World (London:* Hakluyt Society, 1900), Introduction, pp. xxiv ff., and B. Altaner, *Die Dominikanermissionen des 13. Jahrhunderts* (Breslauer Studien zur historischen Theologie; Habelschwerdt, 1924), pp. 116 ff.; for the principal source of information for this episode see Vincent de Beauvais, *Speculum Historiale* (Douai: B. Belleri, 1624), VI, chaps. xxxi ff.

12. For the events as narrated by Franciscan sources see Luke Wadding, *Annales Minorum* (3rd ed.; Quaracchi-Florence: Collegio San Bonaventura, 1931), II, 29-34.

13. *Ibid.*, VI, 399-407. See also *Sinica Franciscana*, I, 424-39, and C. Golubovich, *Biblioteca bio-bibliografica della Terrasanta e dell' Oriente francescano* (Quaracchi-Florence: Collegio San Bonaventura, 1906-27), III, 211-13.

14. See his *Historia Mongalorum,* especially chap. vi, in *Sinica Fran-. ciscana,* I, 76 ff.

15. See O. Franke, *Geschichte des Chinesischen Reiches* (Berlin: W. de Gruyter and Co.), IV (1948), 430 ff., and for the sources and bibliographical notes, *ibid.,* V (1952), 210 ff.

16. V. Bartold, *Turkestan Down to the Mongol Invasion* (Gibb Memorial Series; Oxford and London: G. Lusac, 1928).

17. This episode is narrated only in the Zelada version of his book, see Moule and Pelliot, *Marco Polo: Description,* Vol. II.

18. Casey A. Wood and F. M. Fyfe, trans. and eds., *The Art of Falconry, being the De arte venandi cum avibus of Frederick II of Hohenstaufen* (Stanford, Calif. : Stanford University Press, 1943). This book includes a complete bibliography of the subject. See also Brunetto Latini'a first vernacular description of hunting birds in *Li Livres dou Tresor.* ed. Francis J. Carmody (Berkeley and Los Angeles: University of California Press, 1948), pp. 137 ff.

19. An incomplete list of the animals mentioned or described by Marco Polo is included in J. V. Carus, *Geschichte der Zoologie* (Munich: R. Oldenbourg, 1872), pp. 197 ff.

20. On this legendary "dry tree, " which belongs to the medieval myths of the Earthly Paradise, see the note by Yule, *Book of Ser Marco Polo,* I, 128-39, and Arturo Graf, *Miti, leggende e superstizioni del medio evo* (Turin: G. Chiantore, 1925).

21. Plants mentioned or described by Marco Polo are listed in the standard work of E. Bretschneider, *History of Botanical Discoveries in China* (London: S. Low Marston and Co., 1898), Vol. I, Part I. See also Hermann Fischer, *Mittelalterliche Pflanzenkunde* (Munich: Münchner Drucke, 1929), pp. 44 ff.

22. See Herbert Franke, *Geld und Wirtschaft in China unter der Mongolherrschaft* (Leipzig: O. Harrassowitz, 1949), and Lien-sheng Yang, *Money and Credit in China* (Cambridge, Mass. : Harvard-Yenching Institute, 1952).

23. For the precious stones mentioned above see B. Laufer, *Sino-Iranica* (Chicago: Field Museum of Natural History, 1919). See also Otto Maenchen-Helfen, "Diamonds in China, " *Journal of the American Oriental Society,* LXX (1950), 187, and Edward H. Schafer, "The Pearl Fisheries of Ho-p'u, " *ibid.,* LXXII (1952), 155 ff.

24. See B. Laufer, "Asbestos and Salamander, " *T'oung Pao,* VI (1905), 299-377. The towel could not be delivered to the pope because the papal throne was vacant at the homecoming of the elder Polos from China in 1269.

25. See Franke, *Geld und Wirtschaft,* pp. 119 ff.

26. On these commodities see *Chau Ju-kua: His Work on the Chinese and Arab Trade in the Twelfth and Thirteenth Centuries, Entitled Chu-fan-chih,* trans. and annot. F. Hirth and W. W. Rockhill (St. Petersburg: Printing Office of the Imperial Academy of Sciences, 1911).

27. See Pierre Hoang, "Exposé du commerce public du sel, " *Variétés*

Sinologiques (Shanghai), No. 15 (1898), and Herbert F. Schurmann, *Economic Structure of the Yüan Dynasty* (Harvard-Yenching Institute Studies, Vol. XVI [Cambridge, Mass.: Harvard University Press, 1956]), pp. 165 ff.

28. An incomplete but reliable translation of this interesting treatise by Esson M. Gale (Leiden: E. J. Brill, 1931).

29. See C. Bauer, "Venezianische Salzhandelspolitik bis zum Ende des XIV. Jahrhunderts, " *Vierteljahresschrift für Sozial und Wirtschafts- geschichte*, XXIII (1930), 1 ff.

30. Nevertheless Friar William of Rubruck, who crossed the Asiatic continent up to the capital of the Mongolian Empire in 1253-54, was very much impressed by the great revenue that the Mongol rulers of western Asia and Russia derived from the salt monopoly. See Rockhill, *Journey of William of Rubruck*, pp. 52, 92.

31. For more details and extensive bibliography cf. my book *L'Asia di Marco Polo* (Biblioteca Storica Sansoni, Vol. XXX [Florence: G. C. Sansoni, 1957]), and the forthcoming English translation by John A. Scott, M.A. (Oxon.), University of California Press, Berkeley, California.

E. W. Strong: Hypotheses Non Fingo

1. *Sir Isaac Newton's Mathematical Principles of Natural Philosophy and His System of the World*, trans. Andrew Motte, 1720; trans. rev. and appendix supplied by Florian Cajori (Berkeley: University of California Press, 1947), pp. 546-47.

2. *Ibid.*, Appendix, p. 671.

3. *Ibid.*

4. *Ibid.*

5. *Philosophical Transactions of the Royal Society,* No. 80 (1672), pp. 3075-87.

6. Cited in L. T. More, *Isaac Newton, a Biography* (London, 1934), p. 83.

7. *Philosophical Transactions Abridged* (5th ed. ; London: J. Lowthorp, 1749), I, 147; *Philosophical Transactions*, No. 84 (1672), p. 4091.

8. *Philosophical Transactions*, No. 85 (1672), p. 5004. Pardies' second letter dated May 21, 1672, and Newton's reply from Cambridge on June 2, 1672, appear in this same number of the *Philosophical Transactions*, pp. 5012-13 and pp. 5114-18, respectively. Oldenburg asked Newton for the "proper experiments." Writing from Cambridge on September 21, 1672, Newton responded as follows: "To comply with your intimation of communicating experiments proper for determining the Quaeries which the letter contained, I drew up a Series of such Experiments in design to reduce the theory of colours to Propositions & prove each Proposition from one or more of those Experiments by the existence of common notions set down in the form of Definitions and

Axioms in imitation of the method by which Mathematicians are wont
to prove their doctrines. And that occasioned my suspension of an answer, in hopes my next should have contained the finished said designe.
But before I finished, falling to some other business of which I have
had my hands full, I was obliged to lay it aside & now know not when
I shall take it again into consideration. However if the answer to Mr.
Hooks Consideration will conduce to the determination of those Quaeries
(as in some particulars I think it will) you may if you think fit, publish it; To which end I desire you to mitigate any expressions that seem
harsh, that its publication as you intimated may be done to common
satisfaction. " (MS copy, Keynes collection, Kings College, Cambridge).
Newton's "design" was consummated in his *Opticks* (London, 1704).

   9. *Philosophical Transactions*, No. 85 (1672), p. 5005.
   10. *Ibid.*, pp. 5012-13.
   11. *Isaaci Newtoni opera quae exstant omnia. Commentariis illustrabat
Samuel Horsley* (London, 1770-85), IV (1782), 314-15. Lowthorp *(Philosophical Transactions Abridged*, I, 149) translates only the first and
last sentences of the Latin text, omitting the intermediate portion. Newton concludes with the following comment: "As to the Reverend Father's
calling our Doctrine an Hypothesis, I believe it proceeded from nothing
else, but that he used the Word which first occurred to him, for a Cus-
tom has prevailed, that whatever is explained in Philosophy is called
an Hypothesis. And I had no other Design in shewing my Dislike to that
Word, than to obviate an Appelation, which may mislead those who desire to philosophize in the right Way" *(Ibid.*, p. 151).
   12. L. T. More *(Isaac Newton, a Biography*, p. 78) takes what Newton said at face value in holding that Newton "discarded hypothesis from
science altogether. " Evidence is not supplied by More in support of
his contention that, in Newton's opinion, "all hypotheses ultimately introduced occult forces or substances. " This was not Newton's view of
"mechanical hypotheses. " Moreover, Newton uses the term "hypothesis" in the *Principia* in the sense of "conditions supposed, " e. g., Book
II, sec. ix, "The circular motion of fluids, " begins with the announcement of "Hypothesis: The resistance arising from the want of lubricity
in the parts of a fluid is, other things being equal, proportional to the
velocity with which the parts of the fluid are separated from one another. " Newton subsequently refers to this hypothesis as having been proposed "for the sake of demonstration" (Book II, prop. lii, theorem xl,
Scholium). Such usage accords with the method announced in Book II,
prop. lxix, theorem xxix, Scholium: "In mathematics we are to investi-
gate the quantities of forces with their proportions consequent upon any
conditions supposed; then, when we enter upon physics, we compare those
proportions with the phenomena of Nature, that we may know what conditions of those forces answer to the several kinds of attractive bodies.
And this preparation being made, we argue more safely concerning the
physical species, causes, and proportions of the forces. "
   13. *Opera omnia,*, IV, 324.

14. *Ibid.*, pp. 324-25.

15. *Philosophical Transactions*, No. 97 (1673), p. 6109.

16. In this letter to Oldenburg, Newton is not deviating from his declaration that it is beside his purpose "to examine how Colours may be explained Hypothetically. " Replying to Hooke on July 11, 1672, Newton had said that he made of the corpuscularian hypothesis "at most but a very plausible consequence of the doctrine and not a fundamental supposition" and had so indicated in using the word "perhaps" in introducing it. He is not, however, speaking of the corpuscular composition of light when, after referring to replies made to Pardies and Hooke "concerning the application of all hypothesis to my Theory, " he produces a general rule as concerns "my Hypothesis. " Objectors to the *theory* had persisted in designating it as a hypothesis. If it be so termed, in what does the hypothesis consist? It does not consist of a supposition by which to explain the theory, but of the theory itself as experimentally confirmed. There is, then, as Newton concludes, no basis for objection to the theory of heterogeneal composition on the ground of "my hypothesis" where the latter is nothing other than the experimentally confirmed "original diversities" (MS copy, Keynes collection, Kings College, Cambridge).

17. *Newton's Mathematical Principles*, Appendix, p. 673.

18. *Opera omnia*, IV, 342.

19. *Philosophical Transactions Abridged*, I, 159.

20. *Opera omnia*, IV, 385-94.

21. *Ibid.*, p. 322.

22. MS, Keynes collection, Kings College, Cambridge.

23. Hooke wrote to Newton about the controversy that had arisen. The letter is dated 1675/6 and seems clearly to be in response to Newton's letter of December 21 sent to Oldenburg. "The hearing of a letter of yours read last week in the meeting of ye Royal Society made me suspect y^t you might have been some way or other misinformed concerning me. . . . I have therefore taken the freedom wch I hope may be allowed in philosophicall matters to acquaint you of myself, first that I do noeways approve of contension or finding and proving in print, and shall be very unwillingly drawn to such kinds of warr. Next that I have a mind very desirous of and very ready to imbrace any truth that shall be discovered though it may t(h)wart and contradict any opinions or notions I have formerly embraced. . . . Thirdly that I do justly value yor excellent Disquisitions. . . . Your Designes and myne I suppose aim both at the same thing wch is the Discovery of truth. . . . If therefore you will please to correspond about such matters by private letter I shall very gladly imbrace it. . . . This way of contending I believe the more philosophicall of the two . . . " (MS, Keynes collection, Kings College, Cambridge).

Newton replied (Cambridge, February 5, 1675/6) as follows: "At the reading of yor letter I was exceedingly pleased and satisfied wth yor generous freedome, and think you have done what becomes a true Phil-

osophical spirit. There is nothing wch I desire to avoyde in Matters of
Philosophy more than contention, nor any contention more than one in
print; and therefore I gladly embrace yor proposal of a private corres-
pondance. What's done before many wittnesses is seldome wthout some
further concern then that of truth; but what passes between friends in
private, usually deserve ye name of consultation rather than contention,
and so I hope it will prove between you and me. . . . What DesCartes
did was a good step. You have added much several ways and especially
in considering ye colours of thin plates. If I have seen further it is by
standing upon ye shoulders of Giants. But I make no question you have
divers very considerable experiments besides those you have published
and some of its very probably ye same wth some of those in my late
papers. Two at least there are wch I know you have observed; the di-
lation of ye coloured rings by ye obliquation of ye eye and ye apparition
of a black spot at ye contact of two convex glasses and at ye top of a
water bubble; and it's probable there may be more, besides others w^{ch}
I have not made; so that I have reason to defer as much, or more, in
this respect to you as you would do to me . . . " (MS 141, Keynes collection, Kings College, Cambridge).

24. *Newton's Mathematical Principles*, p. 550.
25. *Ibid.*, p. 400.
26. *Opticks* (London, 1704), p. 1.
27. *Opticks* (London, 1730), p. 377.
28. *Ibid.*, p. 344.
29. *Ibid.*, p. 380.

Otto Struve: The First Stellar Parallax Determination

1. See for example, W. Baade in the *Transactions of the International Astronomical Union*, ed. P. Th. Oosterhoff (Cambridge, Eng.: Cambridge University Press, 1954), VIII, 397; or O. Struve, *Sky and Telescope*, XII (1953), 203.
2. *Memoirs of the Royal Astronomical Society*, XII (1841), 453.
3. *Astronomicheskii Zhurnal*, XXIX (1952), 597.
4. This quotation from Miss Clerke's book may have been introduced by the translator, V. Serafimov. The English text on page 45 of Miss Clerke's *A Popular History of Astronomy during the Nineteenth Century* (2nd ed.; Edinburgh: A. and C. Black, 1887) reads as follows: ". . . Struve made a similar, and somewhat earlier trial with the bright gem of the Lyre, whose Arabic title of the 'Falling Eagle' survives as a time-worn remnant in 'Vega.'"
5. The original English text in Arthur Berry, *A Short History of Astronomy* (London: J. Murray, 1898), p. 363, reads as follows: "Early in 1839 Thomas Henderson (1798-1844) announced a parallax of nearly 1" for the bright star α Centauri which he had observed at the Cape, and in the following year Friedrich Georg Wilhelm Struve (1793-1864) ob-

tained from observations made at Pulkova a parallax of 1/4" for Vega; later work has reduced these numbers to 3/4" and 1/10" respectively."

6. The following is a translation of the relevant sections in Parenago's book: "The discussion of (Bessel's) first series of measurements, from July, 1837, until October, 1838, gave for the parallax of 61 Cygni the value 0".314 ± 0".020. . . . After the instrument was cleaned Bessel's assistant Schlüter undertook a second series of observations, which, together with the first, gave for the parallax a value of 0".348. It should be noted that the mean of several modern series of observations give for the parallax of 61 Cygni, 0".299 ± 0".003. . . . As often happens in the history of astronomy (and also of other sciences) some of the greatest discoveries quickly follow one another. So in this case, in 1839 Henderson found from measurements of the declinations of α Centauri and α Canis Majoris parallaxes of 0".98 and 0".31 (the modern values are 0".755 and 0".376), while W. Struve found for the parallax of α Lyrae 0".261, from measurements of the bright star with reference to its optical companion. Of these three determinations the last turned out to have the greatest error, since contemporary results indicate that the parallax is 0".121 ± 0".004."

7. The Observatory, XLV (1922), 341.

8. W. M. Smart, Foundations of Astronomy (London: Longmans, Green and Co., 1942).

9. Astrophysical Journal, XCVI (1942), 159.

10. R. Wolf, Geschichte der Astronomie (Munich: R. Oldenbourg, 1877), p. 543 (my translation).

11. Memoirs of the Royal Astronomical Society, XI (1839), 61.

12. F. Becker, Geschichte der Astronomie (Bonn: Universitäts-Verlag, 1947), p. 66 (my translation).

13. E. Lebon, Histoire abrégée de l'astronomie (Paris: Gauthier-Villars, 1899), p. 92 (my translation).

14. W. Brunner, Die Welt der Sterne (Zurich: Büchergilde Gutenberg, 1947).

15. P. Couderc, The Expansion of the Universe, trans. J. B. Sidgwick (New York: Macmillan Co., 1952), p. 49.

16. E. A. Fath, The Elements of Astronomy (New York: McGraw-Hill Book Co., 1934), p. 242.

17. E. Zinner, Geschichte der Sternkunde (Berlin: Julius Springer, 1931), p. 526.

18. R. L. Waterfield, A Hundred Years of Astronomy (London: Duckworth, 1938), p. 19.

19. R. A. Sampson, History of the Royal Astronomical Society 1820-1920 (London: Wheldon and Wesley, 1923), p. 89.

20. This postwar edition was revised and partly rewritten by several German astronomers, including W. Becker, R. Müller, H. Schneller, L. Biermann, J. Dick, M. Grotrian, C. Hoffmeister, H. von Klüber, H. Ludendorff, N. Richter, H. Siedentopf, and P. Wellmann. See page 608 (my translation).

21. The value given was actually recomputed by G. A. F. Peters, see *Recueil des mémoires présentés à l'Académie des Sciences par les astronomes de Poulkova, ou offerts à l'Observatoire central par d'autres astronomes du pays* (St. Petersburg: Académie Impériale des Sciences, 1853-59).

22. *Astronomische Nachrichten*, XVI (1838), 66; see also Bessel's letter to von Busch, dated October 10, 1838, presented to the Royal Academy of Sciences in Berlin; also, *Monthly Notices of the Royal Astronomical Society*, IV, No. 17 (1838).

23. *Memoirs of the Royal Astronomical Society*, XI (1839), 61.

24. *Ibid.*, XII (1840), 1.

25. Addidamentum in F. G. W. Struve, *Stellarum duplicium et multiplicum mensurae micrometricae editas anno 1837 exhibens mensuras Dorpati annis 1837 et 1838 institutas*. Adjecta est disquisitio de parallaxi annua stellae α Lyrae (Conv. Exhib. die 27 September 1839).

26. *The Observatory*, XLV (1922), 348.

27. W. Struve, *Rapport à son excellence*, M. *D'Ouvaroff: Etoiles doubles* (St. Petersburg: Académie des Sciences, 1837), p. 42.

28. *Astronomische Nachrichten*, XVI (1838), 70 (my translation).

29. *Briefwechsel zwischen W. Olbers und F. W. Bessel* (Leipzig: Adolph Erman, 1852), p. 419 (my translation).

30. Otto Struve, *Wilhelm Struve* (Karlsruhe: G. Brauns Hofbuchdruckerei, 1895), p. 38 (my translation).

31. In this connection see R. Main, *Memoirs of the Royal Astronomical Society*, XII (1842), 43: "Struve's parallax of α Lyrae is that which, as Bessel modestly acknowledges, excited his ambition to obtain a similar result."

32. I should like at this point to make a personal remark. The uninformed reader may, at the outset, be prejudiced against an account written by the great-grandson of one of the three contenders for the palm of priority. But it is of historic interest that, among the memories that have been transmitted to me by my family, the brightest is the high recognition accorded to Bessel by his rival. There was never any feud between them, and they maintained the closest bonds of friendship until the death of Bessel in 1846. W. Struve never made any claims other than those which I have quoted, and I believe that some of the more recent claims made on his behalf by others are exaggerated. It is difficult to find fault with those accounts that attribute to Bessel the first determination of a fully convincing stellar parallax. In my personal opinion Struve's discovery of a trace of a real parallax in α Lyrae does not constitute his principal contribution: the latter was in the field of double-star astronomy, and for it he received the Royal Astronomical Society's gold medal in 1826. He and Henderson share with Bessel the credit for having brought about the change of the distance scale mentioned in the beginning of this article.